Computer Modelling of Structural Transformations of Nanopores in Fcc Metals

M.D. Starostenkov, A.V. Markidonov,
P.V. Zakharov, P.Y. Tabakov

Published by **Materials Research Forum LLC**
Millersville, PA 17551, USA

Published as part of the book series
Materials Research Foundations
Volume 63 (2019)
ISSN 2471-8890 (Print)
ISSN 2471-8904 (Online)

Print ISBN 978-1-64490-050-5
ePDF ISBN 978-1-64490-051-2

Distributed worldwide by

Materials Research Forum LLC
105 Springdale Lane
Millersville, PA 17551
USA
http://www.mrforum.com

Printed in the United States of America
10 9 8 7 6 5 4 3 2 1

Table of Contents

Introduction

Modern science pays ever more attention to issues related to the production of electrical and thermal energy, in particular by converting nuclear energy. To increase competitiveness, it is necessary that the new technological platform for nuclear energy meets a number of key requirements, among which safety and economy can be singled out. The fulfilment of these requirements depends primarily on the structural materials of working and designed nuclear reactors. So, for example, to increase the efficiency it is necessary to use higher operating temperatures than those characteristic for light water reactors, and, therefore, the use of many existing materials in the core of the reactor becomes impossible [1].

Irradiation and elevated temperatures during the operation of the reactor contribute to fundamental changes in the microstructure, mechanical properties, and geometry of the material due to swelling, climbing, embrittlement, and so on. It is the radiation-induced phenomena that determine the economic feasibility and safe life of the reactor [2]. On the other hand, a purposeful modification of the properties of solids under the radiation exposure is a powerful tool for creating new materials with predetermined properties.

To date, the problem of creating structural radiation-resistant materials is of fundamental importance, which is due to the wide use of nuclear reactors and the rapid development of the rocket and space industry. When developing a new material, it is necessary to determine not only the optimal combinations of the main components and concentrations of various additives but also the thermomechanical processing conditions in which the material retains its physical and mechanical properties within specified limits for a certain time. This task cannot be solved without understanding the processes that lead to the degradation of the studied properties.

When a solid body is irradiated by a beam of accelerated ions, some of them are reflected from the surface, while the rest penetrates deep into the body of the material, slowing down in it. In this case, the kinetic energy of the ions is wasted by elastic collisions with the nuclei of the atoms of the material and when the electronic subsystem is excited. As a result of the elastic collisions, the atoms of the body can be knocked out of equilibrium positions, and colliding elastically with other atoms create a stream of knocked out atoms, forming a cascade of atomic collisions. Atoms in the cascade consume energy to form point defects. In addition, the resulting strong non-equilibrium temperature fluctuations lead to the creation of so-called post-cascade shock waves [3]. Their occurrence is due to the difference between the time of thermalization of atomic oscillations in some finite region and the time of heat removal from it. As a result of a rapid expansion of the strongly heated region, an almost spherical shock wave is formed.

The appearance of nanoscale energy-explosive discharge regions that generate the shock waves is a common phenomenon for all types of corpuscular irradiation. Nevertheless, this fact is practically not taken into account when studying the behaviour of condensed matter under the conditions of radiation exposure. Therefore, the discovery of new phenomena and processes initiated by the post-cascade shock waves and their orientation toward the formation of unique modified atomic structures is a promising and actual direction of radiation materials science. The effects associated with the propagation and generation of post-cascade shock waves have been termed radiation-dynamic [3]. At the same time, high-speed cooperative atomic displacements take place, the process which is flowing at supersonic speed.

The aim of this study is to reveal the mechanisms of structural transformations of nanopores in fcc metals under the cooperative action of groups of atoms, the mixing of which can be considered as the propagation over a crystal wave.

Due to the fact that the objects of research are of a small size, and the processes under consideration are proceeding at a high speed, the use of the computer simulation method seems to be the most rational method of research in this monograph.

Nowadays computer modelling is the method of research equally recognized as the experimental and theoretical methods. Using a computer model, one can test theoretical developments, explain and forecast phenomena not yet fully described by other methods.

In the present work, the computer simulation was carried out using a molecular dynamics method, an undoubted advantage of which is the possibility of modelling atomic systems at a given temperature or at given atomic velocities.

In this monograph, for the first time, the direct effect of the shock waves, which at the microlevel represent high-speed cooperative atomic displacements, on single vacancies and their various clusters in crystals with an fcc lattice, is considered with the help of the molecular dynamics method. It is shown that under the influence of shock waves a homogeneous nucleation of nanopores is possible, the accumulation of which at the tilt grain boundary can cause its bending. It has been suggested that the impact of shock waves can initiate the migration of grain boundaries. The contribution of the created waves to the process of radiation-stimulated diffusion is considered. This can be both the creation of point defects directly in the wave-front, and the change in the elastic fields of defect formations that activate diffusion. The possibility of crushing, dissolving and moving of nanopores under the influence of the created waves is demonstrated. The bifurcation of latent tracks into individual capillaries is shown for the first time.

The study of structural transformations in metals and alloys under the irradiation with charged particle flows, capable of initiating various processes of atomic rearrangements, is of great interest since the mechanism of energy transfer (elastic or inelastic interaction, ionization) can be purposefully changed by choosing the type and energy of the irradiating particles, that opens wide prospects of the use of radiation influence as the tool of technological processing. The presented results will find application in radiation materials science when developing new materials with specified properties. For example, materials for the future innovative nuclear systems (reactors) of the IV generation, as well as for improving the properties of already known materials exposed to various extreme effects during operation. The solution of the given problems will allow us to introduce new, more efficient methods of metal processing. The considered processes of pore formation and structural transformations of nanopores may be used in the development of new methods for controlling swelling. The phenomenon of fission of cylindrical nanopores can be used to create new filters, detectors and cooling elements in nano-electronics. The conducted experiments can serve as a basis for the development of a number of mathematical models. In addition, it is possible to use the results of computer modelling as a demonstrator material that reflects the processes taking place in crystalline structures and is useful for students mastering the course of solid state physics.

1.   Cooperative effects arising in solids under the radiation exposure

1.1   Radiation-dynamic effects

The development of a cascade of atomic collisions, until the moment at which the Maxwellian distribution of colliding particles is established in terms of velocities, occurs over a time of the order of 10^{-12} s. The typical value of the radius of a dense cascade, generated by an ion or a recoil atom with an energy $E > 10$ KeV, is ~ 5 nm, and the time necessary for removing the heat from a cascade region of this size is at least 10^{-11} s for metals [4]. Thus, the thermalization of dense cascades leads to the formation of nanosized regions in which thermal energy is localized [5]. Calculations show that the temperature in this region, which has an ellipsoidal shape with a fairly sharp boundary, can reach the melting point of the material, and the upper limit of the pressure is several units and even tens of GPa [7]. On the boundary separating the cascade region and the unperturbed medium, there is a sharp jump in thermodynamic parameters: a shock wave propagating in a material with supersonic velocity [8] (for the formation of such waves there exists the term *post-cascade shock wave* [3]). Because of the mechanisms of focusing of atomic collisions, initially, the spherical wave is transformed into fragments of the plane waves propagating along close-packed crystallographic directions [9-10].

Theoretical studies of dense cascades, which are characterized by a short particle path length, are carried out in the approximation of a continuous medium since it is senseless to assign to each individual particle its characteristics when the particles collide frequently [8]. The change in the thermodynamic parameters at the late stage of cascade evolution is described by a standard system of hydrodynamic equations, which has a solution in the form of a spherical shock wave [11-12]. The existing estimates show that the pressure at the wave front reaches tens of GPa, while the front width is ~ 1 nm. This high pressure causes a number of interesting effects: the flow of a defect-free material by mixing the atoms of the medium, anomalous mass transfer, diffusionless processes, phase transformations, a sharp increase in the number of displacements per atom in the body of the material, even in the absence of the temperature necessary for the onset of diffusion [11-13].

The propagation of a post-cascade shock wave in a metastable medium when the local energy is minimum allows it to overcome the energy barrier separating the medium from the state of the global minimum. As a result of overcoming the barrier, additional energy is released. If this energy is released at a rate exceeding the rate of dissipation of the wave energy, which is propagating in a metastable medium, then such a wave becomes self-propagating [3]. As calculations show, the wave becomes undamped if the energy of the particle released during damping in a dense cascade exceeds the energy necessary to overcome the energy barrier separating the stable and metastable state.

The term *radiation-dynamic effects* is introduced to denote various shock-wave processes caused by material irradiation, processes of structural rearrangements such as chain reactions initiated by elastic waves, and also phase transformations described by hydrodynamic equations [3]. Thus, these phenomena are separated from radiation-stimulated processes caused by the formation of radiation defects in the irradiated material. The role of radiation-dynamic effects in many cases is crucial when the structure and properties of a solid body subjected to irradiation change.

Experimental studies of radiation-dynamic effects can be found in [3, 7, 14-15]. These works describe the fast-proceeding structural-phase transformations in industrial aluminium alloys after irradiation with accelerated ion beams and are considered as metastable media with increased stored energy. It has been established that due to radiation-dynamic effects, it is possible to improve the intermetallic composition, as well as increase the level of mechanical properties of aluminium rolled products by the use of radiation annealing, which may be an alternative to furnace annealing. In this case, it was possible to reduce the annealing time by one to two orders, and the energy intensity of the process by two to three times. The observed gradual removal of the crystallographic texture of rolling with an increase in the irradiation dose is associated with the explosive rearrangement of the dislocation structure initiated by the generation and propagation of undamped shock cascade waves. It is shown that the structural changes occurring under the influence of irradiation are of a nonthermal nature.

Thus, the processes of formation of shock waves as a result of the evolution of dense cascades of atomic collisions lead to the emergence of a whole series of most interesting phenomena, observed, in particular, at depths considerably exceeding the estimated ranges of the incident ions. A radiation-dynamic effect is the main reason for the initiation of structural-phase transformations in the metastable media.

1.2   Processes of pore formation in solids under irradiation

The most urgent task of radiation materials science is the development of recommendations to combat the swelling of the material since this process is one of the main reasons for the failure of structures operated under irradiation conditions. The radiation swelling takes place because of the development of radiation porosity due to the decomposition of the supersaturated solution of vacancies in the metal, that is, the vacancies formed during irradiation of the material are combined into volumetric clusters representing vacancy pores. Obviously, such amalgamation of defects leads to an increase in the volume of the solid.

The swelling effect caused by the pores with dimensions of about 10 nm was first discovered experimentally in 1967 [16], in the study of austenitic stainless steels. It turned out that swelling of steel can exceed 6% of the initial volume.

It is believed that stationary growth of pores is possible due to the fact that dislocations and dislocation loops, which are an integral structural component of irradiated materials, interact more strongly with interstitial atoms, rather than with vacancies, because of their greater mobility in the crystal lattice [17]. Despite the fact that this difference is only 1 - 2% [8], a stream of displaced atoms flows to the dislocated material in the irradiated material. Because of the excess, the vacancies form various clusters, in particular, the pores. Subsequently, the vacancy pores, which are natural sinks of defects, account for a larger flow of vacancies than interstitial atoms, which is the reason for their growth. In the absence of factors inhibiting the nucleation and growth of pores, the material swells.

Radiation pores are divided into two types: heterogeneous, which are formed on various crystal defects, and homogeneous, arising from spontaneous vacancy clusters. For the growth of the pore, it is necessary to have a sufficiently high vacancy supersaturation, the mobility of vacancies, and stability of the vacancy cluster. The pores are formed practically in all metallic materials during irradiation in the temperature range of $(0.3-0.5)$ T_m.

Calculations based on the theory of elastic continuous media have shown that the tetrahedron of packing defects is the most stable configuration of small vacancy clusters (less than 1000 vacancies) for Au, Ag, Cu, and stainless steel, while for Al, partial or complete dislocation loops of the subtraction type for clusters of smaller or larger size, respectively [18]. In Ni, for clusters consisting of fewer than 200 vacancies, the most energetically favourable configuration is the tetrahedral of packing defects. An increase in the number of vacancies from 300 to 5000 increases the tendency to form partial loops. For the metals named, the dislocation loops are the most stable configuration of large vacancy clusters.

A comparison of the surface energies of a spherical and flat vacancion complex, which is a split loop of dislocations, has shown that for a small number of vacancies the spherical cavity has a lower energy than a dislocation loop, but starting at some critical number of vacancies, the cavity becomes energetically less stable and may collapse with formation of a dislocation loop [19]. The spherical cavity can be stabilized by helium atoms, which are products of nuclear reactions, or local tensile stresses. The compressive stresses, on the contrary, reduce the stability of the spherical cavity.

In general, it should be noted that the calculated data on the morphology of the most stable vacancy clusters, as a rule, diverge from the experimental results, since the observed morphology of clusters in a particular material depends significantly on the presence of impurities that are not taken into account in the calculation.

The final stage of the diffusion evolution of the ensemble of vacancy pores is maturation or coalescence, the driving force of which is the tendency to decrease the free surface of the pores [20-21]. The physical mechanism underlying this process is the thermally activated evaporation of vacancies by pores, as this takes place the small pores evaporate more intensively than the larger ones, and eventually dissolve. Coalescence of pores is most noticeable in cases where the pores are at a sufficient distance from the grain boundaries of the crystals. With prolonged exposures, only large pores located in the centre are observed in crystals, and a nonporous crust is formed along the boundaries [22].

The mechanism of heat-induced coalescence considered above describes well the late stage of the diffusion decay of solid solutions at high temperatures, but it cannot explain the decrease in pore density at lower temperatures. In this case, a radiation-induced mechanism is proposed [23], according to which coalescence is a consequence of the preference for absorption by the pores of intrinsic interstitial atoms, rather than individual vacancies, due to a stronger elastic interaction. Small pores absorb interstitial atoms more intensively and as a result dissolve, while large pores absorb excess vacancies and grow.

The main regularities of the swelling effect were established experimentally. For example, it turned out that the swelling depends to a considerable extent on the temperature at which the sample is irradiated [17]. The curve of the swelling as a percentage of the temperature of the sample is bell-shaped, the maximum being observed at a temperature equal to $(0.4-0.45)\,T_m$. However, as the study shows, the curve for some metals and alloys may have two local maxima at high irradiation doses. Moreover, the swelling of the material at the second maximum can be much greater than at the first.

The degree of swelling of the material depends on its structure and chemical composition. In addition, the swelling level is directly dependent on the presence of mechanical stresses in the sample at the time of irradiation. This factor is important since the structural units of power plants are always under the influence of any mechanical stresses. The change in the value of the stresses from zero to the yield point is accompanied by an almost linear increase in the radiation swelling. Consequently, metal samples under mechanical stress swell faster than samples that are in an unstressed state [24].

A number of well-known papers are devoted to the problem of swelling of the material [17, 25-27]. A study was made of the influence on the formation of a radiation pore of such factors as radiation temperature, irradiation intensity, fluence, the presence of structural imperfections, the presence of gases and non-gaseous impurities. The accumulated extensive experimental material allowed us to construct models of this phenomenon in order to form the basis of the theory. Among the theoretical models, it is necessary to single out the model of diffusion-deformation instability [28-29], according to which the excess vacancies are a cause of elastic tensile stresses. The decrease in Gibbs energy, when these stresses are taken into account, leads to the appearance of the phenomena of ascending diffusion, and, as a consequence, the formation of pores. This model is designed to overcome the difficulties encountered in describing the initial stage of pore formation, as well as explain why the vacancy satiety is realized in the form of pores, rather than vacancy loops.

Theoretical approaches are applicable mainly for homogeneous materials. The presence of impurities and the isolation of secondary phases, which, as we know, can lead to a reduction in the swelling of the material, make calculations much more difficult, and especially for solid solutions, since the evolution of their defect structure is determined both by diffusion of intrinsic point defects and impurity atoms [30] . At present, the study of the radiation swelling of solid solutions is of the greatest interest, since the most obvious way of creating new radiation-resistant materials is the modification of metals by impurities.

Due to the fact that the radiation swelling of the material, in addition to bending and deforming, can lead to welding of individual parts, wedging and overheating of rubbing parts, this phenomenon can be extremely undesirable. Therefore, methods for suppressing the swelling of materials have been developed. These include methods for changing the content of the main components in alloys, alloying of metals, changing the initial microstructure of the material, etc. [24]. In [31-32], a model was proposed that describes the effect of low concentrations of the alloying element on the amount of radiation swelling. A significant containment of radiation swelling is possible even at low doses of the alloying element. The authors showed that in the case of a subdivision impurity, that is, when the atoms of the alloying component are smaller than the atoms of the basic crystal matrix, the ascending diffusion of vacancies is a suppressed ascending diffusion of the atoms of the alloying component. In addition, it has been shown that the oversized impurity also reduces the swelling effect. Another possible way to extend the life of structural elements of nuclear reactors is to partially restore the mechanical properties of austenitic steel by removing radiation damage products from irradiated structures [33-35].

All of the above indicates the need for a systematic study of the processes of pore formation, kinetics of pores, the specific mechanisms of their development in crystalline and amorphous materials, and so on.

From the above review, it follows that the radiation effect on the structure of a solid body is the cause of a variety of important from a practical point of view effects, among which the cooperative processes play a major role. Therefore, such phenomena must be carefully studied to develop main ideas about their nature.

As mentioned above, when studying the evolution of defective structures in a material subjected to the irradiation, it is necessary to take into account such an important factor as the possibility of the formation of post-cascade shock waves. The influence of these waves on the processes of radiation swelling has not yet been considered.

It should be noted that in modern physical materials science, the pores are viewed from two positions:

1. The pore is an integral component of the structure that determines the origin, properties, and purpose of the material (highly porous structures);

2. The pore is the three-dimensional imperfection of the structure.

There are papers dealing with the propagation of shock waves in porous media [36-37], but a study of the formation and behaviour of the pores has shown that the two approaches in question do not agree with one another [38-39]. Consequently, a systematic study of the processes of pore formation, pore kinetics, the specific mechanisms of their development in crystalline and amorphous materials, and so on is required. But without taking into account their influence on the formation of radiation post-cascade waves, one cannot speak of a fully formed theory.

On the basis of the foregoing, a comprehensive study of the formation of elastic waves, their propagation, and interaction with imperfections in the crystal structure of materials is necessary. Computational support for researchers in the study of the directions in question is provided by computer modelling. At present, this method of investigation is used in all the main problems of radiation material science [40-41].

2.   Computer experiment in condensed matter physics

2.1   Methods of computer experiment on the microscale level

At present, a computer experiment is widely used in problems requiring a large number of numerical calculations or which cannot be solved by simple analytical methods. This method of research is universally recognized and is rapidly developing, the wide dissemination of which contributes to the increase in the computing power and the complexity of the studied processes and phenomena [42].

Conducting research on real materials is often associated with significant difficulties. For example, it is not always possible to control the conditions of an experiment, and it can also be difficult to measure the values of various estimated parameters. All of this leads to an ambiguous interpretation of the results obtained. In this case, the use of computer experiments may also be more preferable than real experiments [42].

A computer experiment is a calculation method in which physical processes are modelled according to a given sequence of physical mechanisms [43]. This method occupies an intermediate position between theory and real experiment. Based on theoretical models, the computer experiment is carried out in the form of numerical calculations. The complexity of the models used can virtually increase unlimitedly with the increasing productivity of computing facilities, as a result of which the simulation results will more and more approach the results of a real experiment. Thus, the computer simulation of physical phenomena and processes allows us to repeatedly duplicate costly real experiments, and, in addition, greatly expand the possibilities of theoretical research. Since the computer experiment cannot exist independently of the constructed theoretical models, as well as real experiments that establish a connection between modelling and reality, this method can be called a link between theory and experiment [44].

In the study of condensed matter by computer simulation, two approaches are distinguished by the choice of the scale level: macroscopic and microscopic [8].

Modelling of macroscopic processes is based on the solution of the equations of the physics of continuous media. The most general form of describing the motion of media is the integral conservation laws, which are valid for both continuous and discontinuous solutions. In addition, for continuous solutions, a differential representation of conservation laws can also be used.

With a microscopic approach to modelling a condensed state, a system of interacting atoms is considered. This approach in the study of materials is applicable in the case when the result depends on elementary events, such as the motion of individual atoms or their aggregate. In this case, the time scales of the simulated processes are much smaller than for real macroscopic processes, and the characteristic spatial scales are usually less than the average distance between atoms. In this case, theoretical models of the physics of continuous media become unacceptable.

In addition, a mesoscopic scale level is intermediate between micro- and macro-levels. With this approach, the methods of continual macroscopic description can be used, but taking into account the real structure of the material, including natural internal boundaries and inhomogeneity.

The computer experiment has a number of advantages in comparison with the actual experiment. First, the researcher can fully control the input variables and boundary conditions. This makes it possible to exclude from consideration the influence of an individual physical factor on the result obtained, or to consider its influence on the studied process separately or in combination with other factors. Secondly, it is possible to comprehensively examine the details of the process under investigation, without including additional external influences, which is not achieved by other methods [45].

Carrying out research by means of computer experiments assumes the performance of the following sequence of actions [46]:

1. Definition of the physical model, which is the object of research. In addition to the system of interacting particles, physical models can be dislocations, grains, cracks, and so on;

2. Selection of the parameters of the considered physical model, and determination of the limits in which they can vary;

3. Choice of the mathematical apparatus for formalizing the chosen model;

4. Formalization of the model and selection of criteria allowing to verify the reliability of the obtained numerical results;

5. Approximation of equations, determination of approximation parameters and compilation of computational software for computers;

6. Conducting a numerical experiment and obtaining results;

7. Statistical processing of the results;

8. Verify that the results match the physical criteria of the model;

9. Comparison of the obtained numerical values of the model parameters with the results of real experiments.

When conducting a computer experiment, it is necessary to maintain a certain balance. The physical model should be designed in such a way that all the key physical features of the process or phenomenon under investigation are reproduced, but it should be simple enough, otherwise, it will not be possible to perform the numerical calculations [42].

The choice of the mathematical apparatus of the model depends on the method used in carrying out the computer experiment. At present, in the modelling of condensed matter, three methods can be distinguished [46]: the Monte Carlo method, the variational method, and the molecular dynamics method. This division of modelling methods at the micro-level is conditional, since a rigorous computer experiment, as a rule, is a combination of several of the above methods [42].

The Monte Carlo method refers to a class of stochastic methods that have an advantage over other methods when the physical system under investigation has a large number of degrees of freedom, and also are used in the case of modelling processes that have a long duration in real time. In calculations performed by the Monte Carlo method, the sequence of elementary acts of interchange of particle locations is generated in accordance with the matrix of probabilities of conditional transitions determined by the chosen interparticle interaction potential. The result of these calculations, as a rule, is the average value of some variable or the distribution of its possible values. The Monte Carlo method is widely used to determine equilibrium configurations in alloys and liquids.

The variational method consists in determining the configuration of the system of atoms, at which the potential energy, which is a function of their coordinates, reaches a minimum value. Calculations carried out using this method make it possible to determine static atomic configurations in a stable equilibrium position near the point or lattice defects that have a relatively small volume. The limited use of this method consists in the impossibility of studying the processes occurring at different temperatures since the kinetic component of the total energy of the system is assumed to be zero.

The molecular dynamics method is designed to investigate the motion of individual particles, which are regarded as material points located in the field of forces. For the set of particles under consideration, a system of ordinary differential equations of Newton's dynamics is constructed, for which the Cauchy problem is solved. In this case, the initial and boundary conditions are determined by the physical model in question.

During the experiments that formed the basis of the monograph, the computer simulation was carried out using the molecular dynamics method. The choice of this method is due to the possibility of modelling the system at a given temperature or at given atomic velocities. In addition, there is the possibility of describing the dynamics of the process under study in real time, so that the molecular dynamics method favourably differs from other methods of computer simulation that can be used in the physics of condensed media. Therefore, this method should be considered in more detail.

2.2   Method of molecular dynamics

The method of molecular dynamics was first described in [47], where a three-dimensional one-component system of 256 particles ("hard spheres") was investigated. For several years, it has been developing intensively and is now successfully used to calculate the thermodynamic properties, the structure, temporal correlation functions, and a number of relaxation processes for a wide variety of systems described by different interparticle interaction potentials and having different aggregate states [42]. The method of molecular dynamics is most widely used in the study of defects in crystals [48]. With its help, it is possible to determine the structure, energy, and stresses of various defects, such as vacancies, interstitial atoms, dislocations, packing defects, grain boundaries, and so on [40-41, 46, 49-53]. In addition, the study of such processes in solids as plastic deformation, destruction, diffusion, and others has become very popular with the help of molecular dynamics modelling [54-60]. It is worth noting the possibility of studying of the phase transformations, as well as the processes occurring on the surface of solids, which makes it possible to study the structure and properties of nanoparticles and clusters [61-66].

The method of molecular dynamics is based on a mathematical model consisting of a system of differential equations, a difference scheme, the potential of interatomic interaction, initial and boundary conditions [67]. The method is based on the model representation of a polyatomic system represented as a collection of N material points, the motion of which is described by the classical Newton equations, where each of them:

1. has a mass m_i, a radius vector $\vec{r}_i$ and a velocity $\vec{v}_i$, where $i = 1, 2, ..., N$;

2. interacts with other material points by means of forces $\vec{f}_i = -\dfrac{\partial U(\vec{r}_1, ..., \vec{r}_N)}{\partial \vec{r}_i}$, where

    $U(\vec{r}_1, ..., \vec{r}_N)$ is the potential energy of interaction of the system;

3. interacts with external fields by means of forces $\vec{f}_i^{\,*}$.

In this case, any change in the model will be determined by the system of $2N$ ordinary differential equations of motion:

$$\begin{cases} m_i \dfrac{d\vec{v}_i}{dt} = \vec{f}_i + \vec{f}_i^{\,*}, \\[2mm] \dfrac{d\vec{r}_i}{dt} = \vec{v}_i, \end{cases} \quad i = 1, 2, ..., N. \tag{2.1}$$

Thus, the force $\vec{F}_i$ acting on the particle i, can be represented as the sum of the forces due to the interaction of the particles with each other and the external forces:

$$\vec{F}_i = -\sum_{j=1 (i \neq j)}^{N} \frac{d}{dr_{ij}} \varphi(r_{ij}) \cdot \frac{d\vec{r}_{ij}}{dr_{ij}} + \vec{f}_i^{\,*}, \quad i = 1, 2, ..., N, \tag{2.2}$$

where $\vec{r}_{ij} = \vec{r}_j - \vec{r}_i$, $r_{ij} = |\vec{r}_{ij}|$, $\varphi(r_{ij})$ is the interparticle interaction potential, the choice of which depends on the material considered.

Computer Modelling of Structural Transformations of Nanopores in Fcc Met.   Materials Research Forum LLC
Materials Research Foundations **63** (2019)   https://doi.org/10.21741/9781644900512

To integrate this system, it is necessary to know the coordinates and velocities of all the particles at the initial instant of time. Setting the initial conditions is a non-trivial task since the position of the particles and their velocities have a significant effect on the results obtained in the simulation process. If the simulated medium is a crystal, the preparation of the initial conditions occurs in three stages [61]:

The value of the lattice constant a is calculated at a temperature T for a sample having an infinite length along all the directions and being in equilibrium at zero pressure. For this purpose, a calculated cell with $\tilde{a} = 1$ is constructed, and since the potential energy of each particle is a function of the distance to the surrounding particles $u_i = u(r_{i1}, \ldots, r_{ij}, \ldots, r_{iN})$, $r_{ij} = \left| \vec{r}_i - \vec{r}_j \right|$, $i \neq j$,), then from the virial theorem for a particle with coordinates $(0, 0, 0)$, the following equation can be written:

$$\frac{1}{2} \sum_{j=1}^{N} r_{0j} \frac{\partial u(r_{01}, \ldots, r_{0N})}{\partial r_{0j}} = 3k_B T, \quad r_{0j} = a\tilde{r}_{0j}, \tag{2.3}$$

where $\tilde{r}_{0j} = \sqrt{\tilde{x}_{0j}^2 + \tilde{y}_{0j}^2 + \tilde{z}_{0j}^2}$ is the dimensionless distance from the origin to the j-th particle, k_B is the Boltzmann constant. This equation determines the dependence of the lattice constant a on temperature T;

1. The coordinates of the particles are set by determining the positions of the atoms at the lattice sites which correspond to the minimum potential energy. In the future, at this stage, a random deviation of the particles from the equilibrium position or the kinetic energy of the system, corresponding to twice the temperature $2T$, can be specified. The subsequent relaxation of the model leads to equalization of the kinetic and potential energies;

2. The particle velocities are set in accordance with the Maxwell distribution, which corresponds to the temperature T or $2T$, depending on whether the deviation from the equilibrium position was given in the previous stage, while the total impulse of the system must be equal to zero.

After these steps are completed, the procedure of relaxation of the simulated system must come to a stationary state.

The system (2.1) is a system of ordinary differential equations, for which it is sufficient to know the initial conditions. However, when the modelling objects represented as a particle system, additional requirements may arise, and in order to fulfil them, it is required to indicate the special conditions on the boundaries of the object [67]. Thus, when studying microobjects within a small region of a crystal, the results obtained need to be transferred to a macro volume in order for them to be plausible, that is, it is required to specify the conditions for joining the considered microvolume with the external volume of the crystal, which will be the boundary condition for the studied system [42].

The hard boundary conditions are the simplest, under which the coordinates of the boundary atoms are fixed. This type of boundary conditions is used in the study of point defects when the interaction of atoms located near the defect region with boundary atoms is insignificant [46].

In the case of simulation of defects with large linear dimensions, the mobile type of boundary conditions is used. To implement these conditions, the region under consideration is surrounded by a boundary layer of atoms, which move in accordance with the theory of continuous media [43]. Another way of specifying the mobile boundary conditions is that some pressure is applied to the boundary atoms, which keeps the crystallite from destruction. Sometimes, together with pressure, a viscous force is applied, simulating the outflow of energy from the computational domain to the environment [42].

If it is required to model the interaction with a part of the crystal that is not a part of the computational domain, then it is necessary to use the periodic boundary conditions. Let the computational domain be a cube with an edge length L and include N particles. Then, to create the periodic boundary conditions, the computational cube is surrounded by the same cubes with the same number of particles whose coordinates differ from the coordinates of the particles of the computational cube by the value of L [68]. In this case, if one of the atoms goes beyond the face of the computational cube, the same atom is injected from the opposite side. It should be noted that when using this type of boundary conditions, the maximum range of the interparticle interaction potential should not exceed half the length of the edge of the computational cube, otherwise the atom might interact with its own image [66]. In those cases when the surface of a crystal is modelled, and free from any external influences, the use of boundary conditions is not required.

The main drawback of the method of molecular dynamics is a very high computational cost in obtaining the results. This circumstance leads to an increase in the requirements for the organization of the computing process.

Each iteration in the integration of the system of equations (2.1) can be divided into two stages [67]:

1) Calculation of the interparticle interaction potential and the sum of the forces acting on each atom from the other atoms, and also, if necessary, external forces;

2) Calculation of the new coordinates and velocities of atoms.

The first stage takes most of the computational time. One possible way to increase the computation speed is to introduce the cut-off radius r_c of the potential. This assumption is justified by the fact that for most materials the interparticle interaction potential is short-range, that is, the interaction of atoms at a distance $r > r_c$ can be neglected. After the introduction of the cut-off radius, the calculation time of the potential energy and forces is significantly reduced, but in this case, a singularity appears in the cut-off point, if the potential does not approach zero smoothly. Therefore, it is necessary to check the effect of the cut-off radius on the characteristics of the system under consideration [42].

To accelerate the search for atoms that interact with the atom under consideration, a block method can be used [57]. In this case, the entire crystal is divided into elementary cubic blocks, the length of the edges of which is equal to the cut-off radius of the potential. The blocks are numbered in order, and the coordinates of the atoms of the crystal determine the numbers of the corresponding blocks. In order to determine the forces acting on a certain atom, only atoms located in the same or a contiguous block are considered.

In the second stage, it is necessary to integrate a large number of equations with given initial and boundary conditions, since it is required to determine the trajectories of a set of particles. The numerical solution of the system (2.1) leads to errors in determining the particle trajectories. Therefore, regardless of the choice of the numerical solution method, it is necessary to determine the time step of the difference scheme, and also to control the rounding errors and their subsequent correction [46].

When determining the length of the time step Δt, an empirical rule is used, which makes it possible to achieve the necessary accuracy of calculations for a small time [69]. This rule can be analytically expressed in the form

$$\Delta t \approx \Delta x / \vartheta, \qquad\qquad (2.4)$$

where Δx is the characteristic mean free path of the atom, and ϑ is the characteristic velocity.

In order to control round-off errors, the law of conservation of energy is most often checked [42]. In addition, it is possible to calculate the temperature and compare it with known theoretical values.

Among the many known methods of integrating equations (2.1), the Werle algorithm [70] became very popular, the use of which is very convenient if there are no non-conservative forces in the system (2.1), although there are schemes that also take the friction into account [67]. The idea of he algorithm is to decompose the radius vector of the particle at time instants of

$\vec{r}(t + \Delta t)$ and $\vec{r}(t - \Delta t)$ in the Taylor series to the third power in Δt:

$$\vec{r}(t+\Delta t)=\vec{r}(t)+\frac{d\vec{r}(t)}{dt}\Delta t+\frac{1}{2}\frac{d^2\vec{r}(t)}{dt^2}\Delta t^2+\frac{1}{6}\frac{d^3\vec{r}(t)}{dt^3}\Delta t^3+O(\Delta t^4)$$
, (2.5)

$$\vec{r}(t-\Delta t)=\vec{r}(t)-\frac{d\vec{r}(t)}{dt}\Delta t+\frac{1}{2}\frac{d^2\vec{r}(t)}{dt^2}\Delta t^2-\frac{1}{6}\frac{d^3\vec{r}(t)}{dt^3}\Delta t^3+O(\Delta t^4)$$
. (2.6)

Adding (2.5) and (2.6), and taking into account that the second derivative of the radius vector with respect to time is the acceleration $\vec{a}(t)$, we obtain

$$\vec{r}(t+\Delta t)=2\vec{r}(t)-\vec{r}(t-\Delta t)+\vec{a}(t)\Delta t^2+O(\Delta t^4)$$
. (2.7)

Thus, the position of the particle is determined from the two previous positions, and the acceleration can be calculated through the forces acting on it.

In the Werle algorithm, there is no need to calculate the particle velocities, but if this is required, for example, to calculate the kinetic energy, then it can be obtained by subtracting equation (2.6) from (2.5):

$$\vec{v}(t)=\frac{d\vec{r}(t)}{dt}=\frac{1}{2}\Delta t[\vec{r}(t+\Delta t)-\vec{r}(t-\Delta t)]+O(\Delta t^2)$$
(2.8).

In addition to the basic kinematic relations, the method of molecular dynamics also uses the auxiliary relationships that make it possible to calculate various characteristics of the system [42]. Thus, the potential energy of the system consists of the potential energy of the pair interaction of atoms and the potential energy due to the action of external forces:

$$U(\vec{r}_1,...,\vec{r}_N)=\frac{1}{2}\sum_{i=1}^{N}\sum_{j=1(i\neq j)}^{N}\varphi(r_{ij})+\sum_{k}\varphi_k(r_k),$$ (2.9)

where k are the indices of the atoms to which the external forces $\vec{f}_k^*$, described by the potential $\varphi_k(r_k)$, r_k are the magnitude of the deviation of the k-th atom from the equilibrium position. The kinetic energy of the system is determined by the following formula

$$E = \frac{1}{2}\sum_{i=1}^{N} m_i |\vec{v}_i|^2 .$$
(2.10)

The temperature, which is an important characteristic of the system, can be expressed in terms of the average kinetic energy of the particles:

$$T = \frac{2\langle E \rangle}{s N k_B},$$
(2.11)

where s is the number of degrees of freedom of the system.

The calculation of the mean values of the components of the stress tensor is done using the following expression

$$\sigma^{\alpha\beta} = -\frac{1}{V_n}\sum_{i=1}^{n}\left(m_i v_i^{\alpha} v_i^{\beta} + \frac{1}{2}\sum_{i \neq j}^{n} F_{ij}^{\alpha} r_{ij}^{\beta} \right),$$
(2.12)

where α and β are indices of the tensor component, V_n is the volume of the cell by which averaging is performed, n is the number of particles in the considered cell.

The method of integrating the system of equations (2.1), considered above, ensures the conservation of the energy of the system. In addition, it is assumed that the volume and number of particles also remain constant. In this case, a micro-canonical system is modelled. Sometimes it becomes necessary to model a canonical or isothermal-isobaric ensemble. To maintain thermodynamic quantities at a constant level, there are four methods: differential, proportional, integral and stochastic [48].

In the differential method, the thermodynamic quantity f has a fixed value, and there are no fluctuations about the mean value $<f>$.

When using the proportional method, all values associated with f at each integration step are adjusted by a correction factor that sets the given value of the thermodynamic value f. This correction factor determines the magnitude of the fluctuations around $<f>$.

The essence of the integral method consists in expanding the Hamiltonian of the system by including new independent quantities, reflecting the effect of the external system which fixes the state of the desired ensemble. The evolution of these quantities over time is described by the equations of motion obtained from the extended Hamiltonian.

In the stochastic method, the values associated with the thermodynamic quantity f are assigned values in accordance with the modified motion equations, in which some degrees of freedom are additionally changed stochastically to give the desired value $<f>$.

2.3   Potentials of the interparticle interaction

The most important procedure, when using the molecular dynamics method, is the choice of the interatomic interaction potential. From this choice depends not only realistic modelling but also the speed of the calculation, keeping in mind the use of resource-intensive algorithms. Historically, the solution to this problem has been divided into two areas: the "first principles" method and the empirical potential method [67].

The "first principles" method is based on the solution of the Schrödinger equation for a condensed medium, which is considered as a set of interacting nuclei and electrons. Thus, to calculate the structure and properties of a solid body, it is necessary to specify the initial location of the nuclei, solve the quantum problem of the location of a given number of electrons in their field, then calculate the free energy of such a system, and then, by moving the nuclei, it is required to minimize this energy [4]. Obviously, without some simplifying approximations, the solution to such a problem proves to be very difficult.

Using the "first principles" method, the energy characteristics of point defects, packing defects [71-72], twins [72-73], antiphase boundaries in ordered alloys [74] and grain boundaries [75-78] were calculated, but the computational complexity of this method limits its wide use. Therefore, in modelling in the physics of condensed media, the empirical potentials that are simpler from a computational point of view have become more popular. Nevertheless, the conclusions derived from the quantum-mechanical theory are often used in the construction of such potentials.

The construction of the empirical potential can be divided into two successive stages [67]: the choice of the analytical form of the potential and the selection of the parameters of the potential.

At the first stage, the choice can be made based on the quantum-mechanical theory, physics or chemistry of interatomic bonds. In this case, the potential can include several functions that depend on interatomic distances, coordination numbers, and so on.

At the second stage, it is necessary to select the parameters of the functions included in the potential. In this case, the selection is carried out by fitting to the known physical characteristics of the substance obtained experimentally or on the basis of quantum mechanical calculations, which can be divided into the following groups.

The first group is the structural characteristics. This group includes the most commonly used constant in the selection, - the lattice period of the considered crystal. The interatomic distance is determined experimentally with the help of X-ray diffraction analysis and electron micro-diffraction methods with an accuracy of 0.01 - 0.1% [4].

The second group is the power characteristics. These parameters characterize the "rigidity" of interatomic interactions. This group includes such popular constants as elastic moduli. The accuracy of measuring the elastic moduli for single crystals is up to 0.1-1%, and for polycrystals, it is no better than 1% [4].

The third group is the energy characteristics. This group includes, for example, the vacancy energy and the activation energy of its migration. It is very difficult to determine the energy characteristics of a material experimentally. The fact is that the absolute values of such parameters do not exist in the physical nature of these characteristics. They are dependent on, for example, the actual state of the material or the domain of the energy parameter within the material structure. Therefore, depending on the method used, the accuracy can lie in the range of 2 - 20% [4].

Thus, when the empirical potentials are used, and when they are adapted to the three types of parameters described above, a fairly accurate study of the structural and force changes in the material is possible [79]. Moreover, many changes occurring in the material at the atomic level, mainly depend on the "hardness" of the interatomic bonds, that is, they are determined by the force parameters. But in calculating the energy parameters, it is necessary to take into account only their relative changes and consider the direction of these changes in the study of structural-phase transformations in materials, and especially in the case of nanocrystals and low-dimensional systems.

The first potentials used to describe the condensed media were developed on the basis of a pair interaction model, the basic idea of which is quite simple. Any state of the crystal can be expressed in terms of the coordinates of the atoms. Then the energy can be represented as a function of the distances between all the atoms, that is, the interaction of two atoms will depend only on their mutual position, while the position of the other atoms has no effect. Thus, the energy of the crystal will be equal to the sum of the energies determined by the sums of the pair interactions of the atoms:

$$U(\vec{r}_1,...,\vec{r}_N) = \sum_{i=1}^{N} u(\vec{r}_1,...,\vec{r}_N)_i = \frac{1}{2}\sum_{i=1}^{N}\sum_{j=1(i\neq j)}^{N}\varphi(r_{ij}). \qquad (2.13)$$

In general, the representation of energy in the form of a sum of pair interactions has no quantum-mechanical justification and is possible only for the crystals of inert gases in which the interatomic bonds are determined by van der Waals forces. In metals, in addition to the direct ion-ion interaction between atoms, there is also an indirect interaction, through the effects of electron distribution. For example, the proof of the non-pairiness of the interaction is the violation of the Cauchy relation in metals [4].

In [80-81], an alternative approach was proposed to describe the energy states in compounds with a metal bond, called the submerged atom method. This method is based on the quantum-mechanical theory of the electron density functional, according to which the contribution to the energy of arbitrarily located nuclei from the interaction with electrons can be represented as a single-valued functional of the total electron density. It is assumed that the state and energy of an atom is determined only by the density of electrons, and the density itself in a metal is represented as a linear superposition of the contributions of individual atoms. In addition, the electron density produced by one atom is spherically symmetric [67]. The potential energy of the crystal, in this case, will be represented as the sum of the energy of pair interaction of atoms and the energy of interaction of atoms with an electron gas:

$$U(\vec{r}_1,...,\vec{r}_N) = \frac{1}{2}\sum_{i=1}^{N}\sum_{j=1(i\neq j)}^{N}\varphi(r_{ij}) + \sum_{i=1}^{N}F(\rho_i), \tag{2.14}$$

where $F(\rho_i)$ is the energy of introduction of the i-th atom into the electron density, ρ_i is the total electron density for the i-th atom.

The density ρ_i is created by spherically symmetric one-electron density functions $f(r_{ij})$ of other atoms:

$$\rho_i = \sum_{j=1(i\neq j)}^{N}f(r_{ij}). \tag{2.15}$$

The parameters of the functions $\varphi(r_{ij})$, $F(\rho_i)$ and $f(r_{ij})$ can be determined in two ways [48]:

1. Calculation based on the quantum-mechanical theory, that is, the theory of the electron density functional. However, it was not possible to obtain exact potentials, based only on this method;

2. Selection of parameters in such a way as to obtain a set of properties of the crystal that coincide with known experimental values of physical quantities (lattice parameter, crystal bond energy, elastic constants and vacancy formation energy).

Often, the function $\varphi(r_{ij})$ is given in the form of some pair potential or an n-degree polynomial, $f(r_{ij})$ is determined from the quantum-mechanical representations, and the function $F(\rho_i)$ can be obtained from the equation of state [67].

The potential calculated by the IAM method, when compared with the pair potential, takes into account multiparticle interaction due to the embedded function $F(\rho_i)$.

Due to this term, the potential correctly describes the decrease in the binding energy per bond, as the coordination number increases [48].

2.4   The methodology of computer experiments

As already mentioned above, the computer experiments that form the basis of this monograph, were carried out using the molecular dynamics method, which is by far the most powerful tool for studying the microscopic structure of crystals. This method favourably differs from others in the ability to solve problems associated with structural-energy transformations of materials under conditions of temperature-force effects [8].

To describe the interatomic interaction in computer experiments, the Johnson potential calculated in the framework of the immersed atom method was used [82]. In this potential, the functional form of the electron density $f(r)$, the pair potential $\varphi(r)$, and the embedded energy $F(\rho)$ are given in the following way:

$$f(r) = f_e \exp\left[-\beta\left(\frac{r}{r_e}-1\right)\right], \quad r \le r_c$$

$$\tag{2.16}$$

$$\varphi(r) = \varphi_e \exp\left[-\gamma\left(\frac{r}{r_e}-1\right)\right], \quad r \le r_c, \tag{2.17}$$

$$F(\rho) = -E_c\left[1-\frac{\alpha}{\beta}\ln\left(\frac{\rho}{\rho_e}\right)\right]\left(\frac{\rho}{\rho_e}\right)^{\alpha/\beta} - \Phi_e\left(\frac{\rho}{\rho_e}\right)^{\gamma/\beta}, \tag{2.18}$$

where $\alpha = 3\sqrt{\Omega B / E_c}$, Ω is the atomic volume, B is the bulk compression modulus, E_c is the binding energy, $\rho_e = 12 f_e$, $\Phi_e = 6\varphi_e$, r_e is the shortest equilibrium distance between atoms, r_c is the cut-off radius of the potential. The value $f_e = S(E_c / \Omega)$ is the scaling constant, which can be taken equal to one for pure metals [83].

The function $f(r)$ is a positive decreasing function, and $F(\rho)$ is a function with positive curvature, therefore, as the density increases, the interaction becomes more repulsive [48]. The potential parameters $\varphi_e, \beta, \gamma, r_c$ are determined by fitting to the lattice parameter a_0 or to the atomic volume Ω, the binding energy E_c, the vacancy formation energy E_v, the bulk compression modulus B and the shear modulus G. The values of the potential parameters calculated in [84] are given in Table 2.1, and the graphs of the functions (2.16), (2.17) and (2.18) are presented in Figure 2.1.

Table 2.1 Johnson potential parameters

Metal	f_e	φ_e, eV	α	β	γ	r_c, Å
Ni	0.41	0.74	4.98	6.41	8.86	4.84
Au	0.23	0.65	6.37	6.67	8.20	5.70

$$\frac{dT(t)}{dt} = 2\frac{T_0 - T(t)}{\tau},$$

(2.19)

where τ is the characteristic time of interaction with the reservoir.

Thus, the temperature deviation decreases exponentially with the time constant τ. By varying τ, it is possible to change the intensity of heat exchange and adapt it for specific purposes. For example, for initial balancing of the system, τ can be assumed to be rather small, but when modelling the equilibrium system, it must be set sufficiently large. As a rule, during the calculations, a value of the order of 1 ps is assigned to τ.

The change in kinetic energy at each step is modelled by rescaling the velocities of the atoms of the system by multiplying their values by the coefficient λ, determined from the condition [86]

$$\lambda = \sqrt{1 + \frac{2\Delta t}{\tau}\left(\frac{T_0}{T(t - \Delta t)} - 1\right)}.$$

(2.20)

The main drawback of the Berendsen thermostat is the incorrectness of the physical description of small systems, because of the uneven distribution of energy over the degrees of freedom.

To perform the calculations, we used an XMD package of the molecular dynamic modelling [87], developed by John Rifkin at the University of Connecticut (USA). The obvious advantage of this package is a wide range of supported potentials, comparative ease of use and openness of source codes. When using XMD, the interparticle potential is given by a tabular method with the subsequent approximation.

In addition, in carrying out individual computer experiments, the software package developed by the authors was also used [88].

At the final stage of the experiment, it is necessary to perform an analysis of the crystal structure. To do this, a number of visualizers are used. So, for example, RasMol program [89], developed by Roger Sayle at the Department of Biomolecular Structures of Edinburgh University (Great Britain), is used to directly visualize the atoms of the computational cell. This program is distinguished by the simplicity and logical structure of the user interface.

To study dynamic processes, a visualizer of atomic displacements is used. In this case, the displacements of atoms are represented in the form of segments whose length is dependent on a given scale. If the scale of the displacements is equal to one, then the segments join the initial and final positions of the atoms. Thus, the trajectory of the atom's movement during the experiment will be shown in the form of a broken or straight line starting at the point of its position at the start of the experiment. An increase in the displacement scale makes it possible to analyse sufficiently small atomic displacements, which may not even be noticeable when the atomic matrix is directly mapped.

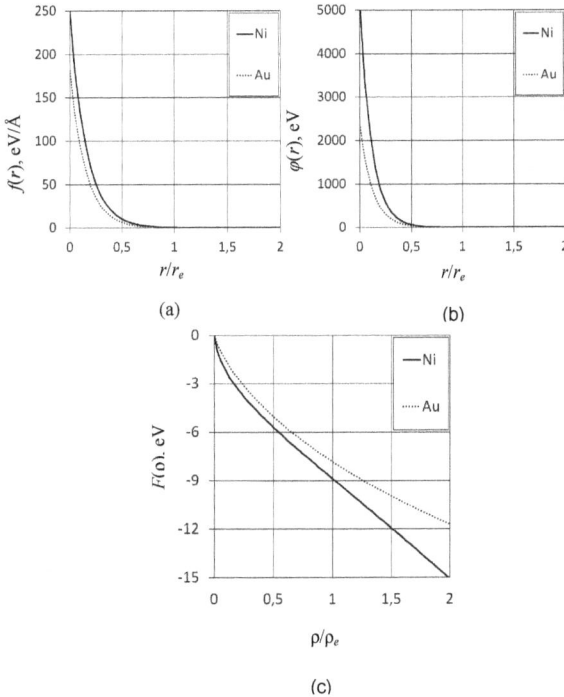

(a)   (b)

(c)

Figure 2.1: Graphs of functions that determine the Johnson potential: (a) function (2.16), (b) function (2.17), (c) function (2.18).

Since the crystal structures containing various defects are considered, a visualizer of the distribution of potential energy is necessary. This visualizer allows us to accurately show the defects in the calculation block, as well as the distribution of the internal local stresses. The essence of the visualizer lies in the fact that for each of the atoms the binding energy is calculated, and in accordance with this value, the atoms are painted over in one or another colour shade.

In addition, a close-packed overlay visualizer consisting of lines connecting atoms in three close-packed directions is used. This visualizer is useful in studying linear defects, restructuring the structure as a result of deformation, crystallization, and so on.

The methods used to implement the project are traditional for computer modelling. Therefore, we can be sure that the use of proven and well-proven research methods ensure accurate results.

3.   Influence of cooperative atomic displacements on the processes of pore formation in crystals

3.1   Modelling of the post-cascade shock wave and the initiation of low-temperature self-diffusion by it

The characteristics of the post-cascade shock wave, which distinguishes it from the waves obtained by other methods, are the large amplitude of atomic displacements, as well as the small front width, which is commensurate with the lattice parameter of the crystal [12]. In addition, the dimensions of the region of the thermal peak, the formation of which is the cause of the generation of this wave, reach several nanometers. Therefore, to create it, a crystallographic plane $\{1\bar{1}0\}$ was chosen in the computational cell, containing the boundary atoms which were assigned a speed equal in magnitude to the velocity of the longitudinal sound waves c_p and whose vector was oriented along the normal to the selected plane. As a result of a subsequent relay and co-operative atomic displacements, a running wave is formed, whose front width does not exceed several interatomic distances, and the amplitude of atomic displacements considerably exceeds the amplitude of thermal oscillations. The distance travelled by the shock wave prior to its degeneration into a sound wave depended on the starting velocity of the atoms and was calculated in tens of angstroms. At the same time, the defects in the computational cell were arranged in such a way that the impact on them was exerted by the shock wave.

Let us perform the calculation of the basic thermodynamic parameters of the generated shock wave. A computer experiment was carried out on a computational cell simulating nickel crystallite and having the form of a parallelepiped. The orientation of the crystallite was set as follows: the X-axis was directed along the crystallographic direction $<1\bar{1}0>$, the Y-axis was along $<11\bar{2}>$, and the Z-axis was $<111>$. As a potential function of the interatomic interaction, the Johnson potential calculated in the framework of the immersed atom method was used [90]. The step of numerical integration of the equations of motion was 5 fs. Surface effects were eliminated using the periodic boundary conditions. Since the use of the above method of generating a shock wave inevitably leads to an increase in the temperature of the computational cell, a proportional thermostat was used to store it. Thus, during the computer experiment, the canonical ensemble was modelled.

With a long experiment, the shock wave can reach the opposite boundary of the computational cell, as a result of which it is possible that the wave passes through the considered region again. To avoid this, in some cases, a viscous force is applied to the atoms of the opposite boundary, simulating the outflow of energy from the computational domain to the environment. To calculate the thermodynamic parameters of the wave, we divide the computational cell into elementary volumes so that after structural relaxation each of the volumes contains one atomic plane $_{\{1\bar{1}0\}}$. Then, after the start of the computer experiment, each volume is analysed. Thus, the obtained distribution of relative parameters along the X axis at the temperature in the computational cell of 600 K and the starting velocity of the boundary atoms 1.5 c_p (for nickel c_p = 5630 m/s [91]) is shown in Fig. 3.1. We note that carrying out a computer experiment at a constant temperature leads to a sufficiently rapid damping of the shock wave. Therefore, it makes no sense to consider the distribution of thermodynamic parameters throughout the entire computational cell, but it suffices to consider a small perturbation region.

In the computer experiments conducted, the pressure induced by the shock front was calculated in GPa. This value largely depends on the number of boundary atoms, which are assigned a velocity when generating a wave. Nevertheless, in spite of the large magnitude of the stresses arising, the simulated crystal lattice remained elastically distorted during the passage of the wave. In addition, there was no observed splitting of the shock wave into an elastic precursor and the subsequent plastic front, and the wave consisted only of an elastic component. Obviously, this is due to the lack of imperfections in the lattice, which may be a nucleus of plastic shear. For example, it was shown in [92, 93] that the ideal crystal lattice of copper remains elastic up to a pressure of 30 GPa, while with the inclusion of point defects, packing defects begin to form even at 5 GPa. In addition, in [94] it is also said that the plastic deformation associated with the formation and motion of dislocation dipoles is achieved at lower temperatures and much smaller uniaxial deformations if the crystal structure contains point defects. The absence of splitting of the shock wave in an ideal crystal during the computer simulation was previously documented in [95]. In connection with the foregoing, the behaviour of the shock wave generated in the conducted experiments and the response of the modelled crystal structure do not contradict the known data.

An additional study showed that if the initial velocity is assigned to atoms of several neighbouring planes $_{\{1\bar{1}0\}}$ during the formation of the wave, then the formation of point defects is possible in the computational cell, which should contribute to the process of self-diffusion.

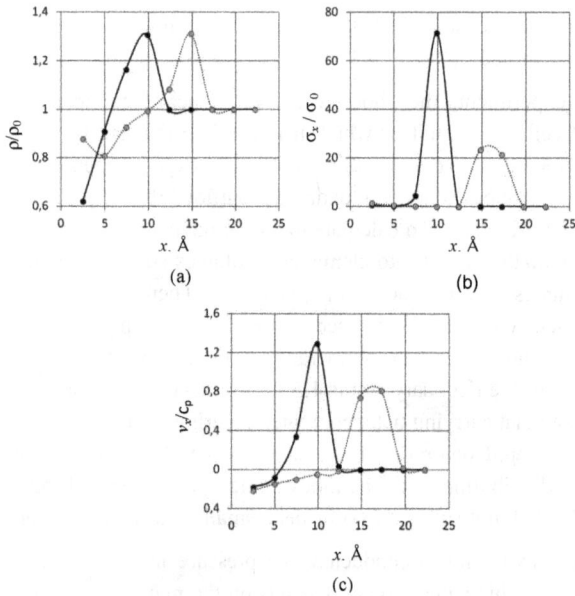

Figure 3.1: *The distribution of the relative thermodynamic parameters along the X-axis when the shock wave passes through 0.05 (●) and 0.1 (◉) ps: (a) density of atoms (ρ_0 is the density in the ideal lattice), (b) normal stress (σ_0 is the internal pressure in the lattice at a given temperature), (c) projection of the velocity (c_p is the velocity of longitudinal sound waves).*

Note that, with the current level of development of the nuclear industry, nuclear power and space technology, the need for radiation-resistant materials that withstand significant mechanical and thermal loads increases. Radiation stability of materials is largely determined by diffusion processes. Under the influence of irradiation, the kinetics of various activated processes in solids can significantly differ from analogous phenomena in the absence of radiation exposure, which is most clearly manifested, for example, in the acceleration of diffusion processes, called radiation-stimulated diffusion. As of now, there is no a single mechanism explaining all the observed behaviour of this process. In the case of threshold radiation effects, the acceleration of diffusion can occur as a result of the "radiation shaking" of a solid body during the relaxation of the metastable states formed [96]. If the contribution of irradiation to the formation of defects by the impact mechanism [97] as well as the formation of long-lived electronic excitations is not taken into account, then, as shown in [98], the radiation stimulation of diffusion processes is associated with the deviation of the temperature dependences for the frequencies of atomic transitions from the equilibrium position based on the Arrhenius law.

When considering the thermal diffusion in metals and alloys, the diffusion flux is determined by the concentration of equilibrium defects. The dominant diffusion mechanism is the vacancy mechanism. In the case of elastic deformations, excessive vacancies can be formed and, in this case, mass transfer depends not only on the number of defects but also on, for example, phase transitions. However, the phase transitions, as a rule, are observed at elevated temperatures, when the role of non-equilibrium defects decreases. The acceleration of diffusion due to the presence of non-equilibrium defects can be investigated at lower temperatures if additional sources of their generation are created. This can be achieved, for example, by generating a shock wave in a solid body, which will be demonstrated later.

First, we check the used model for the reliability of the results obtained. With this in mind, we determine the main characteristics of self-diffusion, among which, as a rule, we consider the migration energy of the defect U_m and the activation of self-diffusion Q, and also the pre-exponential factor D_0 in the Arrhenius equation

$$D = D_0 e^{-Q/kT},$$

(3.1)

where D is the self-diffusion coefficient, k is the Boltzmann constant, and T is the temperature.

The self-diffusion coefficients along the three orthogonal directions x, y and z are calculated by the formulae:

$$D_x = \frac{\frac{1}{N}\sum_{i=1}^{N}(x_{0i} - x_i)^2}{2t}, \ D_y = \frac{\frac{1}{N}\sum_{i=1}^{N}(y_{0i} - y_i)^2}{2t}, \ D_z = \frac{\frac{1}{N}\sum_{i=1}^{N}(z_{0i} - z_i)^2}{2t},$$

(3.2)

where x_{0i}, y_{0i}, z_{0i} are the coordinates of the ith atom at the initial instant of time; x_i, y_i, z_i are the coordinates of the ith atom at time t; N is the number of atoms in the computational cell. The self-diffusion coefficient D was determined as the arithmetic average of D_x, D_y, and D_z [99].

The experiments were carried out on a computational cell containing a single vacancy. The choice of this defect is due to the fact that the vacancy diffusion mechanism is generally accepted as the main one. Based on the results of the experiments, the dependence of $\ln D$ on the reciprocal temperature was constructed. According to (3.1), this dependence is direct, and its slope determines the migration energy (since the defect has already been created):

$$U_m = -k \frac{d}{dT^{-1}} \ln D. \tag{3.3}$$

The pre-exponential factor is defined as $D_0 = \ln D(0) \cdot N$, where $\ln D(0)$ is the value of $\ln D$ at the point of intersection of the straight line with the $1/T$ axis.

The dependence of $\ln D$ $(1/T)$ is shown in Figure 3.2. Note that in order to exclude the contribution to the coefficient of self-diffusion of thermal vibrations of atoms, the duration of the computer experiments was $2 \cdot 10^5$ steps or one nanosecond, and, in addition, after a specified number of steps, the procedure of structural relaxation followed by multiple zeroing of the atomic velocities. Based on the obtained data, the equation of the pairwise linear regression was constructed. The calculated coefficient of determination $R^2 = 0.985$ indicates a strong linear connection.

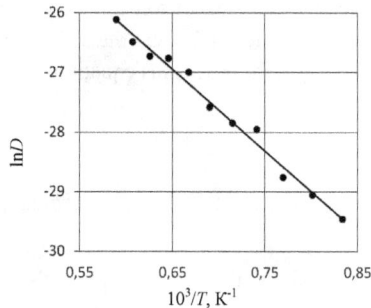

Figure 3.2: *Dependence of the natural logarithm of the self-diffusion coefficient lnD on the reciprocal temperature 1/T, constructed using the considered model.*

The activation energy of self-diffusion is composed of the energy of vacancy migration and the energy of its formation U_v. The energy of formation of a vacancy can be estimated as the difference between the energies of a computational cell with an embedded vacancy and a computational cell consisting of the same number of atoms, but constituting an ideal crystal lattice. The obtained values of the main characteristics of self-diffusion are presented in Table 3.1.

The results of the calculations presented in Table 3.1 indicate the suitability of the model used. Note that the discrepancy D_0 is not critical, since the spread of the values of the pre-exponential factor is observed in many research works, even when calculated by the most reliable isotope methods, and can differ by orders of magnitude [4, 101].

Table 3.1 The main characteristics of self-diffusion calculated in the framework of the considered model.

	Characteristics			
	U_v, eV	U_m, eV	Q, eV	D_0, m^2/s
Model	1.63	1.18	2.81	$3.54 \cdot 10^{-6}$
Literature [4, 91, 100, 101]	1.40 - 1.80	1.04 - 1.30	2.80 - 2.88	$1.30 \cdot 10^{-4}$ - $9.30 \cdot 10^{-4}$

Let us now consider a computational cell that simulates an ideal crystallite in which a shock wave is generated by assigning to a group of boundary atoms a velocity exceeding the sound velocity c_p. To study the effect of the wave on self-diffusion, we calculate the mean square displacements of atoms using formulae (3.2). Note, that in this case, it does not make sense to consider the long-time interval on which the mean square displacement of atoms is determined since it is required to determine the contribution to the self-diffusion of only the propagating wave. The length of the time interval t is set in such a way that the wave having the maximum velocity does not cross the boundary of the computational cell. The effect of thermal vibrations of atoms was eliminated by carrying out a subsequent procedure of multiple zeroing of the velocities of atoms. The coefficient δD, determined during these calculations as the arithmetic mean value of D_x, D_y and D_z, characterizes the process of atomic migration, and is proportional to the self-diffusion coefficient (δ is the dimensionless proportionality coefficient which depends on the size of the computational cell and the duration of the experiment). Since δD is calculated after the passage of the shock wave, we call it the coefficient of forced migration of atoms. The results of the calculations are shown in Fig. 3.3. To approximate the data, we used polynomials of the third degree.

As follows from the analysis of Fig. 3.3, the passage through the computational cell of the shock wave activates elementary acts of self-diffusion, the cause of which are Frankel pairs that originate in the wave front. There are estimates that, in the case of heavy ion irradiation, the pressure in the front of post-cascade shock waves can exceed not only the real but also the theoretical yield strength of solids [3]. In this case, a flow of a defect-free material with the mixing of the atoms of the medium and the formation of defects is possible. Therefore, the cause of activation of self-diffusion, observed in the computer experiments, is fully accomplishable and does not contradict known facts.

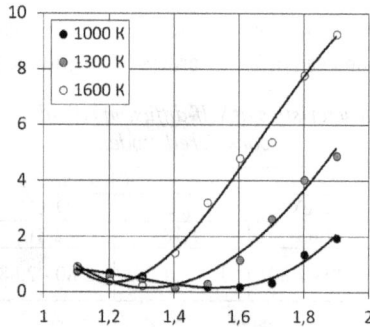

Figure 3.3: *Dependence of the forced migration of atoms δD on the initial velocity v_0 of the group of atoms which generate the shock wave for different starting temperatures of the computational cell.*

As follows from the analysis of Fig. 3.3, the passage through the computational cell of the shock wave activates elementary acts of self-diffusion, the cause of which are Frankel pairs that originate in the wave front. There are estimates that, in the case of heavy ion irradiation, the pressure in the front of post-cascade shock waves can exceed not only the real but also the theoretical yield strength of solids [3]. In this case, a flow of a defect-free material with the mixing of the atoms of the medium and the formation of defects is possible. Therefore, the cause of activation of self-diffusion, observed in the computer experiments, is fully accomplishable and does not contradict known facts.

The coefficient δD reaches a high value even at a wave velocity of 1.1 $\cdot c_p$, which is due to the short duration of the computer experiment. An increase in the time interval t will not reveal the role of the shock wave in activation of the self-diffusion process since in this case, the contribution of thermal processes increases. In this connection, we can conclude that the coefficient δD allows us to characterize more qualitatively the process of self-diffusion.

According to the data presented in Fig. 3.3, it can be seen that the increase in the coefficient of forced migration of atoms is caused not only by an increase in the starting temperature of the computational cell but also by an increase in the wave velocity. In this case, there exists a velocity interval in which an abnormal decrease in the value of the coefficient δD is observed. The reason for this decline is as follows. An increase in the wave velocity leads to the formation of not a single, but multiple interstitial atoms, which are crowdion complexes. The data of formation are much more mobile than single interstitial atoms, and, in addition, the spontaneous recombination radius for such defects is much larger. Therefore, after the passing of the wave front, the crowdion complexes recombine with the vacancies, and if the crowdions moved only along one atomic series, then the displacement data are not considered as the diffusion path in the formulae (3.2). Such a process is similar to the correlation effect for self-diffusion [101]. The subsequent increase in the wave velocity leads to the removal of crowdion complexes from the vacancies, the "link" between them is lost, and recombination is not observed. Therefore, the coefficient of forced migration of atoms begins to grow. We note that at high wave velocities the calculated values of the coefficient δD approach the self-diffusion indices in liquid nickel, which are $2.5 \ 10^{-9}$ - $7 \ 10^{-9}$ m^2/s, depending on the calculation procedure [102].

Thus, the computer experiments have shown that the post-cascade shock wave can be one of the causes of radiation-stimulated diffusion.

Obviously, the analysis of self-diffusion processes on models of only an ideal crystal lattice is not a complete study. Therefore, it is necessary to estimate the role of structural imperfections. Ideally, the linear or volumetric defects, which are sinks for the point defects, are suitable for this. From the position of simplifying the calculations, preference is given to dislocations, namely, screw ones. The choice of a screw dislocation is due to the following considerations. First, virtually any crystal contains a growth dislocation, which is a screw one. Secondly, an edge dislocation creates regions of dilation and, therefore, interaction with point defects is asymmetric. Thus, the effect of the edge dislocation on self-diffusion will depend on the mutual arrangement of point defects and the extra-plane. Obviously, this fact is essential only for computer experiments. A screw dislocation is devoid of this shortcoming. Therefore, we shall carry out an additional investigation of the computational cell simulating a crystal lattice with a screw dislocation.

The screw dislocation was created as follows. In the middle of the computational cell, an imaginary axis was drawn, after which a base plane was allocated from the centre to the edge of the cell, on which the atoms would experience a zero displacement when constructing the dislocation. All the remaining atoms of the computational cell were displaced in the given direction by an amount of

$$d_i = b \cdot \sin(\alpha_i), \tag{3.4}$$

where b is the Burgers vector, α_i is the angle between the reference plane and the atom i. As a result, the atoms are twisted around the given axis. After the manipulations, the structural relaxation of the computational cell was carried out before the system entered the state with minimal energy. To preserve the geometry of the computational cell, whose appearance is shown in Fig. 3.4, a combination of hard and periodic boundary conditions was used.

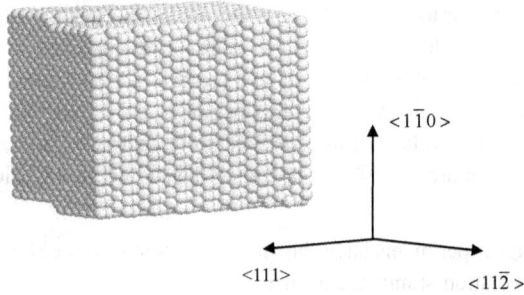

Figure 3.4: *The computational cell containing a screw dislocation with the Burgers vector a/2.*

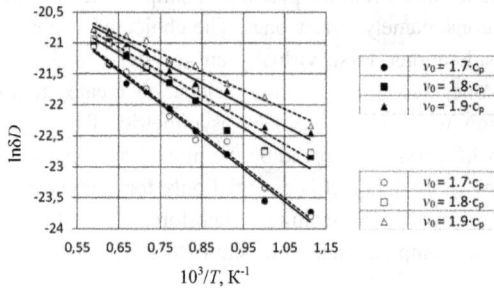

Figure 3.5: *Dependence of the natural logarithm of the coefficient of forced migration of atoms δD on the reciprocal temperature 1/T, calculated using the computational cells that model the ideal lattice (solid regression line) and the lattice with a screw dislocation (dash regression line).*

The conducted study has shown that the forced migration coefficients of atoms, calculated after the passage of the shock wave through this computational cell, exceed the analogous coefficients determined in the cell simulating an ideal crystal lattice. The results of the calculations are shown in Fig. 3.5. The reason for increasing the values of the coefficients δD is as follows. A screw dislocation with the Burgers vector $a/2 <1\bar{1}0>$ in a crystal with an fcc lattice splits into partial dislocations with vectors $a/6 <1\bar{2}1>$ and

$a/6_{<2\bar{1}\bar{1}>}$, forming a packing defect. The interstitial atoms formed as a result of the passage of the shock wave diffuse to this structural imperfection (see Fig. 3.6a), lowering its energy. In the absence of a packaging defect, the Frankel pair can be recombined (see Fig. 3.6b), which has already been discussed above. Since the angular coefficients of the lines constructed in Fig. 3.5 decrease in absolute value as the wave velocity increases, this will lead to a decrease in the calculated values of the defect migration energy.

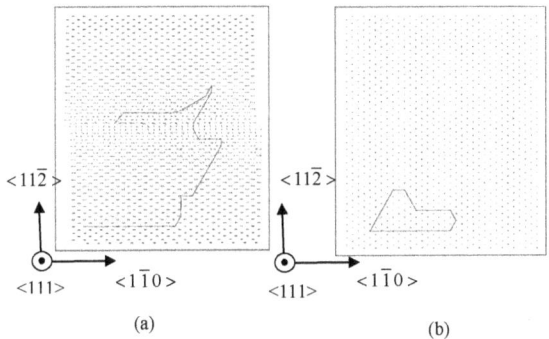

(a)   (b)

Figure 3.6: *Atomic displacements in the computational cells simulating a lattice with a screw dislocation (a) and an ideal crystal lattice (b), after the passing of the shock wave. The initial velocity of the atoms forming the shock wave is $1.1 \cdot c_p$. Starting temperature of the computational cell is 1000 K.*

We note that in [103] the decrease in the values of the migration energy of defects with an increase in the flux of high-energy particles by which the nickel alloy was irradiated was also observed experimentally. In addition, some of the values in Fig. 3.5 were obtained at temperatures insufficient for the thermal activation of self-diffusion by the vacancy mechanism.

Thus, as a result of this study, it was found that the main mechanism of self-diffusion, activated as a result of the passage of shock waves through the crystal structure, is crowdionic. It takes also place at temperatures insufficient for the implementation of the vacancy mechanism. An increase in the wave velocity leads to an increase in the values of the coefficient δD, and makes it commensurate with the coefficients calculated for the liquid nickel. If the crystal lattice contains structural imperfections, the obtained values of the coefficient δD increase due to, for example, attraction and subsequent settling of interstitial atoms on the packing defect.

3.2   Pore formation under the influence of post-cascade shock waves

Of great practical interest is the investigation of the behaviour of a material under the influence of the fluxes of high-energy particles, leading to the formation of point defects. The importance of this study is due to the need to create radiation-resistant structural materials capable of operating under extreme conditions without significantly changing their properties. The dimensional instability, manifested in the form of phenomena of radiation climb or radiation swelling, is caused by the development of a new defect structure as a result of the climbing up of dislocations, as well as, the growth of pores. Pore nucleation is the result of the decay of an unstable system of excess vacancies formed as a result of material irradiation, and its growth is possible due to the asymmetry of interaction of various imperfections of the crystal structure with point defects, and is determined by the presence of locally directed flows of vacancies at the pore. The required ascending diffusion can occur, for example, in the fields of elastic stresses created by bulk radiation defects of the cluster type. The contribution of post-cascade shock waves to the initiation of ascending diffusion was not previously estimated.

Based on a quasi-thermodynamic approach to the description of the formation of a new phase in a condensed medium, the main problem of homogeneous nucleation of the pore was formulated in [104-105] and it can be described as follows. If we regard the pore as a "phase of emptiness," then the formation of the nucleus of such a phase by fluctuation, with a size exceeding a certain critical value determined by the specific surface, is unlikely. To solve this problem, a phenomenological theory has been developed, according to which the combining of vacancies in the pores occurs under the action of elastic tensile stresses, the source of which is the excess vacancies themselves. Obviously, these results can be supplemented if we consider the behaviour of excess vacancies in the region of discharge arising after the passage of the shock front. In this connection, the purpose of the computer experiments presented below was to reveal the contribution of post-cascade shock waves to the upward diffusion of vacancies and to determine the possibility of homogeneous nucleation of the pore under the influence of these waves. During the study, the model described in the previous section was used.

Consider a computational cell containing randomly distributed vacancies. As a result of the passage of waves, there is a directional migration of vacancies to the source of the perturbation. The propagation of a wave is a successive collision of atoms on the principle of a relay, and if there is a vacancy in the atomic series, the neighbouring atom occupies a vacant place. This mechanism of motion of vacancies is realized in the case of the generation of waves in the computational cell at a temperature insufficient for the activation of diffusion processes.

To analyse the spatial distribution of vacancies, it is necessary to introduce some dimensionless characteristic that would allow a visual interpretation. As such, one can use the degree of filling Π, the inverse value of the porosity of the material. To calculate the degree of filling, the number of vacancies in the atomic series is determined, and if they are absent, then Π is taken equal to one, and if there are no atoms in the series, then $\Pi = 0$. As a result, we obtain a matrix, each element of which is equal to the degree of filling of an individual atomic series of the computational cell. After this, an image is constructed where a rectangle, painted in a certain shade of grey, corresponds to each element of the matrix. In limiting cases, when $\Pi = 1$, the rectangle is coloured white, and if $\Pi = 0$, it is black. A similar visualizer is used in the construction of Fig. 3.7 (the calculation of the degree of filling is carried out for atomic series <111>), where the result of the following experiment is presented. The starting random distribution of vacancies in the computational cell is set (see Fig. 3.7a). After the generation of three shock waves, generated at an interval of 10 ps, a procedure of structural relaxation follows by multiple zeroing of the velocities of the atoms and a new configuration of the vacancies is investigated (see Fig. 3.7b). The visualizer shows a significant decrease in the concentration of vacancies and the formation of a pore nucleus near the boundary of the computational cell.

It was shown in [106] that as a result of an annealing of the fcc crystal at temperatures lying in the interval (0.5-1) T_m, the pores are formed in depleted zones with a total concentration of vacancies in the crystal above 25%. In our computer experiments, the pore nucleus, formed under the impact of shock waves, was obtained with a decrease in the concentration of vacancies to 10% and a decrease in the temperature of the calculated cell to $\approx 0.2 \cdot T_m$. We note that if under the given conditions a computer experiment is performed without generation of shock waves, then the pore nucleus is not formed, and the vacancies are rearranged into fragments of tetrahedra of packing defects.

It is known that the pore sizes are limited by the specific surface energy, and after overcoming a certain critical size, the pore collapses with the formation of a dislocation loop, since the spherical shape becomes energetically unfeasible. The pore is stabilized as a result of filling it with helium nuclei, which are products of nuclear reactions, or in the case of compensation of surface energy by local tensile stresses. Obviously, under the influence of such stresses, the pore nuclei can form at a much lower concentration of non-equilibrium vacancies. Thus, when setting the deformation in the computational cell of the 3D strain $\varepsilon = 5\%$ and a temperature of 300 K, the pore nuclei in the computer experiments are formed at 8% of the vacancy concentration. The deformation was modelled by changing the interatomic distances in the computational cell. In the case of generating the shock wave in the computational cell, the pores can be obtained by reducing the concentration of vacancies to 5%.

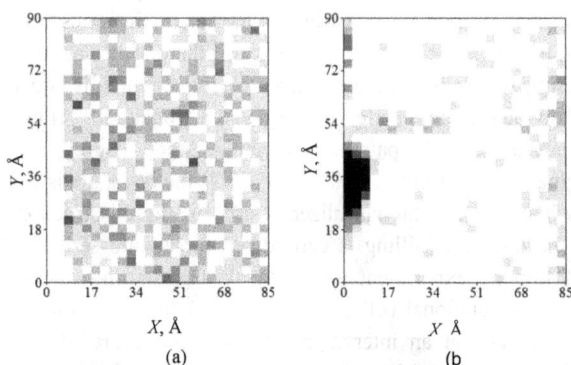

X, Å
(a)

X Å
(b

Figure 3.7: Visualization of the degree of filling Π of the computational cell at the beginning of the experiment (a) and after the generation of three shock waves propagating from left to right with respect to figure (b). The temperature of the calculated cell is 300 K. The white colour corresponds to an ideal crystal lattice, and the black colour corresponds to a pore.

To confirm that a pore is formed after the passage of the shock wave, we shall carry out the following experiment. Let us create two regions with a random distribution of vacancies in the computational cell (see Fig. 3.8a). After the front of the shock wave overcomes one of the regions, a procedure for multiple zeroing of the velocities of atoms is included in order to damp the wave and thereby exclude its effect on the second region. As can be seen from Figs. 3.8b and 3.8c the nuclei of a pore are formed only in the region that the wave has crossed. Thus, the conducted computer experiments demonstrate the possibility of the formation of pore nuclei during the passage of the post-cascade shock wave along the computational cell containing vacancies. In this case, for the nucleation of a pore, the concentration of vacancies can be much lower than in experiments without generation of a shock wave. This is due to the discharge domain formed after the wavefront. The vast majority of metals used in practice are polycrystals.

Their main structural element is the grain boundary, which can completely determine the properties of a nanocrystal. It is the grain boundaries that have a significant effect on the diffusion processes. It is known that the diffusion rate along the boundaries is orders of magnitude higher than inside the grain, and this difference is especially noticeable at low temperatures. The free volume of the boundary separating the grains from each other is equivalent to the existence of a stress of 3D strain [107]. The urgency of the study of pore formation at the grain boundaries is due to the significant effect of pores on the processes of grain-boundary slippage, as well as, the kinetics of grain growth. In connection with this, we will carry out an additional study of the processes of pore formation in the computational cell that simulates the boundary of the tilt grains. To simulate the boundary of the tilt grains, the technique described in [108] was applied. The common-type grain boundary was created by dividing the computational cell into two blocks and rotating them relative to each other by an angle $\theta/2$ along the crystallographic direction <111>.

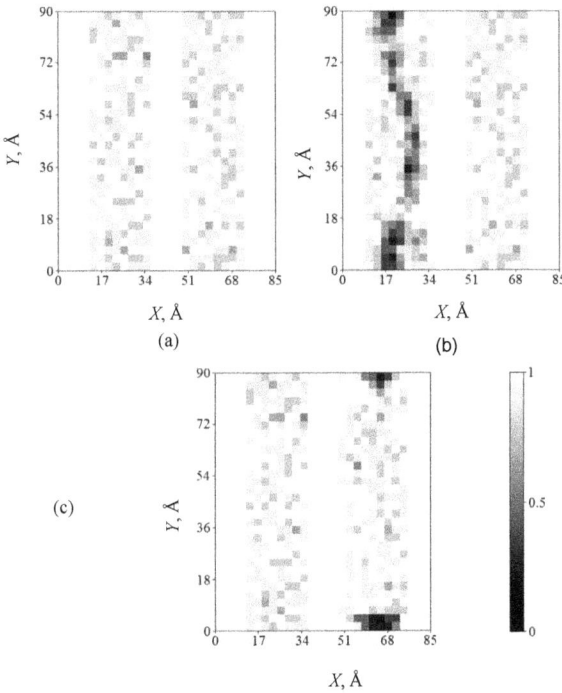

Figure 3.8: Visualization of the degree of filling Π of the computational cell subjected to 3D strain $\varepsilon = 5\%$, at the beginning of the experiment (a) and after generation of the shock wave from left to right (b), relative to the figure, and from right to left (c). The temperature of the cell is 300 K.

The resulting grains approached a distance at which the potential energy of the computational cell was minimal, followed by a relaxation procedure. To preserve the geometry of the computational cell, a combination of hard (in the direction of the X and Y axes) and periodic boundary conditions was used.

The study was carried out on the computational cells containing grain boundaries with misorientation angles $\theta_1 = 9°$ and $\theta_2 = 18°$. According to the classification proposed in [109], in the first case we are talking about a small-angle boundary, and in the second case, about a large-angle boundary. The main characteristics of the simulated boundaries are given in Table 3.2. The specific energy γ, per unit area of he boundary, was calculated as the energy difference of the computational cell containing the grain boundary and the energy of an ideal computational cell consisting of the same number of atoms. The excess volume ΔV was calculated as a linear expansion in the direction perpendicular to the boundary, expressed in fractions of the lattice parameter.

Table 3.2 Structural characteristics of simulated grain boundaries.

Characteristic	Misorientation angle θ		Literary data
	9°	18°	
Specific energy γ, J/m^2	0.648	0.972	0.690 - 0.840 [110]
Excess volume ΔV, a	0.101	0.113	0.094 - 0.125 [111]

Note that the data given in the last column of Table 3.2 takes also into account the boundaries of a special type. Therefore, a direct comparison of the obtained characteristics with these data is not entirely correct, nevertheless, it can be asserted that they do not contradict the known data.

Let us do the following experiment. In one of the grains, we will create a certain number of vacancies distributed randomly, and in the second grain, we will generate a shock wave. We shall study the effect of a wave on the vacancy cluster at a temperature insufficient for thermal activation of diffusion. The computer experiment showed that after the passage of the wave, the vacancies are shifted toward the grain boundary. It is known that the boundary is not an insurmountable obstacle to waves. For example, when a grain boundary of a special type $\Sigma 7$ is intersected, only about 20% of the energy of a solitary wave is scattered in the grain boundary region [112]. In the event that several waves are generated, the vacancies begin to settle on grain-boundary dislocations and subsequently delocalize. For example, Figs. 3.9a and 3.9b show the computational cells with different grain misorientation angles, and with a vacancy concentration of 2%.

After the generation of twelve shock waves, in addition to the vacancies that have settled on the dislocations, a group located in the second grain passes through the dislocation cores (see Fig. 3.9b). We note that when carrying out an experiment with a high-angle grain boundary, the number of vacancies that have moved to the neighbouring grain is much smaller (see Fig. 3.9d). The accumulation of vacancies on grain-boundary dislocations activates a pipe diffusion, which is the main diffusion mechanism at low temperatures. The decrease in the number of vacancies outside the grain boundary region subsequently leads to a decrease in the role of bulk diffusion.

It should be noted that in the computer experiments there was a climbing of dislocations due to absorption of vacancies. This mechanism can be realized in the case of a constant influx of vacancies [113], which in our case provides a series of shock waves.

Let us consider computer experiments in which the computational cells were subjected to 3D strain. As the starting point, the distribution of vacancies shown in Figs. 3.9a and 3.9c was used. After the creation of several shock waves and the subsequent relaxation of the structure, it was observed the displacement of grain boundary dislocations to new equilibrium positions, that is, buckling of the boundary, which occurs as a result of the formation of pores during the drain of vacancies onto dislocations. Such a buckling, as shown in [114], can lead to the disintegration of the entire boundary, if the values of the external stresses are sufficient to significantly remove one of the dislocations from the boundary. The subsequent displacement of the group of freed lattice dislocations causes local plastic deformation and the formation of elongated grain.

Consider the results of the experiment described above in more detail. Fig. 3.10a shows a fragment of the {111} plane of the computational cell containing the grain boundary with $\theta = 18°$. For clarity, the grain boundary dislocations are highlighted in the figure, which are combined in pairs and represent vertex dislocations with the Burgers vector. In the case of a comprehensive deformation of a computational cell with $\varepsilon = 3\%$, the vacancies drain onto one of the dislocations under the impact of shock waves, which leads to the formation of a pore that, as a result of the action of the subsequent series of waves, begins to shift toward the source of the perturbations. Following the impact of waves, the vacancies, that have settled on the dislocation core, start relocating, the extra plane of the dislocation "sprouts" deep into the crystal, which becomes much simpler under tensile conditions, entraining a pair dislocation behind it. Thereafter the dislocations remote from the pores begin to slip to form a vertical wall. This process leads to the enlargement of the right grain due to absorption of the left grain (see Fig. 3.10b).

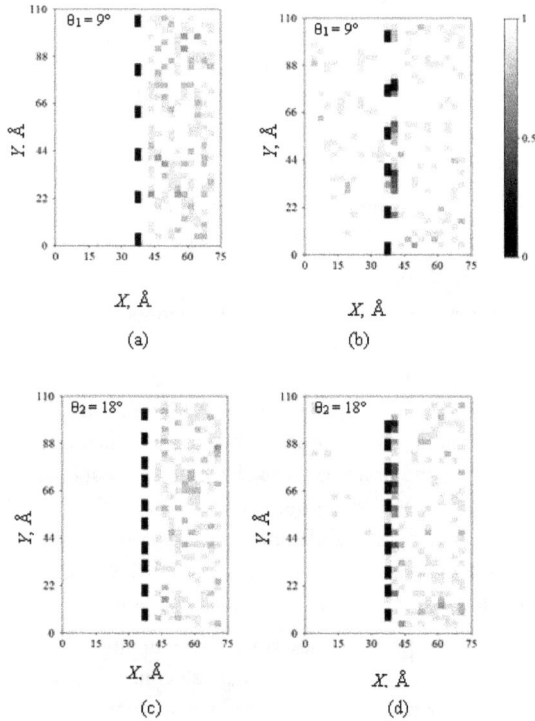

Figure 3.9: *Visualization of the filling degree of the computational cell containing the grain boundary with the misorientation angle θ at the beginning of the experiment (a, c) and after the passing of twelve shock waves generated every 2.5 ps (b, d). The temperature is 300 K. The vertical black elements in the centre of the images correspond to the wall of the nuclei of grain-boundary dislocations. The waves propagate from left to right with respect to the figure.*

Note one feature. The vertex dislocations are located at an angle of $\pi/3$ to the slip plane and is much energy required to shift them. But, as shown in [109], the displacement of such dislocations is a cooperative process involving an entire group of atoms. Moreover, the smaller the misorientation angle θ, the higher the probability of occurrence of this process. Indeed, when carrying out a computer experiment with a computational cell having $\theta = 9°$ and conditions analogous to the experiment described above, a much greater bending of the boundary is observed, due to the displacement of grain boundary dislocations over longer distances. Besides, as shown by additional research, the most intense collective atomic displacements take place near the created pore.

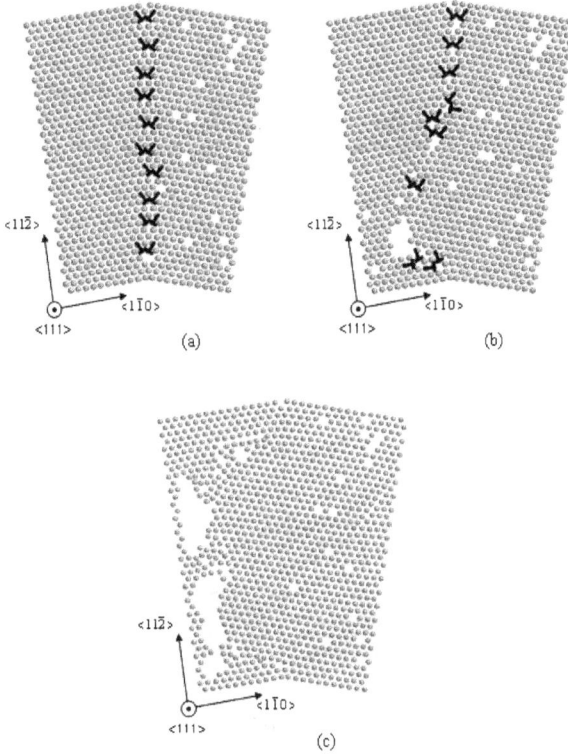

Figure 3.10: *A fragment of the {111} plane of the computational cell containing the grain boundary with an angle of misorientation θ = 18° at the beginning of the experiment (a) and after passing of the ten shock waves generated every 2.5 ps, with a 3% (b) and 4 % (c). The temperature of the cell is 300 K.*

It should be stipulated that during the computer experiment, without creating vacancies or with a decrease in the magnitude of the deformation, the dislocation slip is not observed.

With increasing deformation up to $\varepsilon = 4\%$, the pore is formed at the junction of the grains. In addition, local amorphization of the grain boundary layer is observed near the pore, because of which it is not possible to reveal the nuclei of grain-boundary dislocations. As a result of the subsequent displacement of the pore under the influence of waves, one of the grains also grows (see Fig. 3.10c). With this amount of deformation, the pore formation is observed even without creating vacancies in one of the grains.

The computer experiments presented in this section indicate the possibility of homogeneous pore nucleation in the region of tensile stresses that occurs behind the front of the shock wave. A directional flow of vacancies arising under the impact of a series of waves leads to subsidence of vacancies on natural drains, for example, on grain boundary dislocations. Moreover, the temperature at which this process is observed is insufficient for its thermal activation. The pores are formed under the influence of external tensile stresses in the grain boundary region, the displacement of which by the shock waves causes the growth of grains.

The formation and growth of pores is a characteristic sign of structural changes during radiation impact on a solid body. Therefore, it is necessary to investigate the effect of post-cascade shock waves on the pores in the crystal.

3.3   Structural transformations of nanopores

The change in the structure of structural materials at the micro level under radiation, mechanical and thermal influences is the subject of intensive experimental and theoretical studies, because these changes affect the physical and mechanical properties of materials. The primary structural defects for such extreme effects are point defects. The interaction of these defects with each other, macrodefects of the structure, internal and external stress fields lead to the formation of pores, dislocation loops, phase separations, and so on.

The problem of pore formation under the irradiation of materials has traditionally been given a special attention since these processes are associated with the phenomenon of swelling of the material. It is known that, depending on the size, the pore can remain stable, transform into a tetrahedron of packing defects or into a dislocation loop. For example, it was shown by a computer simulation in [106] that in the computational cell of the fcc crystal, the pores are formed at a vacancy concentration above 25%. In this case, if the radius of the pore is less than three coordination spheres (19 vacancies), then it is reconstructed into a tetrahedron of packing defects. Pores consisting of 43 or more vacancies remain stable at all temperatures considered in the work. In the opinion of the authors of [115], it is energetically feasible to form spherical pores in the fcc lattice for the vacancies with an amount of fewer than 140 units, and for a larger number, clustering in the form of dislocation loops.

The purpose of the computer experiments presented in this section is to investigate the processes of structural transformations of nanopores under the influence of high-speed cooperative atomic displacements, considered as post-cascade shock waves. The relevance of this research is due to the fact that studies of phenomena of vacancy porosity of materials open both ways to combat the problem of swelling of solids and helps to interpret correctly the results obtained in the study of irradiated structural materials used in reactors and accelerators.

The experiments were carried out on a computational cell simulating the crystallite of gold and having the form of a parallelepiped. The orientation of the crystallite was set as follows: the X-axis was directed along the crystallographic direction $<1\bar{1}0>$, the Y-axis was along $<11\bar{2}>$, and the Z-axis $<111>$. To exclude the influence of surface effects, the periodic boundary conditions were used in all directions. To create a pore, a sphere with a certain radius was specified in the crystal structure. Then the centre of the sphere was combined with one of the lattice sites, and all the atoms falling into this sphere were removed.

It should be noted that only a pore formed in an amorphous body can possess a simple spherical shape. In a crystal, the surface energy is a function of the crystallographic direction, and for some syngonies, the differences in its values, depending on the orientation, can reach 50% [38]. Factors affecting the shape of the pore are also the kinetics of its growth and the crystallographic anisotropy of the interaction of vacancies with pores. According to the calculations, the surface energy is minimal for tetradecahedrons in the fcc lattice [39]. Therefore, after removal of the atoms, the structural relaxation of the computational cell was carried out before the system entered a state with a minimum energy.

In the experiments carried out, nanopores of different radii were used, consisting of a different number of vacancies. Thus, Fig. 3.11 shows the images of atoms forming the surface of the pore, that is, the number of nearest neighbours of which is different from twelve.

When constructing the images in Fig. 3.11, the atoms were removed from the computational cell, the binding energy of which differed from the binding energy in the ideal lattice. As follows from the analysis of Fig. 3.11, the pores of the "correct" shape, having a minimum surface energy, consist of fragments of eight planes of the {111} type and six {100} type.

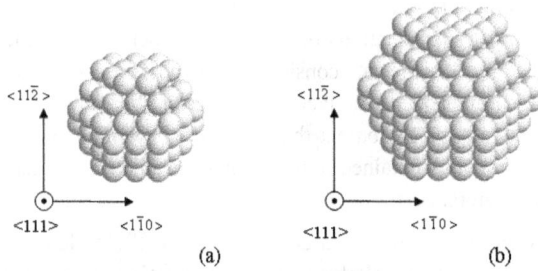

Figure 3.11: The surface of vacant nanopores consisting of 38 (a) and 116 (b) vacancies.

Let us consider the process of structural transformation of nanopores at different temperatures of the computational cell. Computer experiments have shown that the process of restructuring the nanopore into a tetrahedron of packing defects begins at a temperature of ≈ 0.45 Tm (melting point of gold is 1336 K [91]). At a temperature below this, the pores remain stable. In the tetrahedron of packing defects, the four faces represent the packing defects of subtraction in the {111} planes, and the six edges are vertex dislocations with the Burgers vector $a/6$ <110> [116]. Only small pores were reconstructed into ideal tetrahedra, the number of vacancies in which was sufficient for constructing an ideal tetrahedron. Thus, the pore, consisting of 15 vacancies, was reconstructed into an ideal tetrahedron at a temperature of 900 K (see Fig. 3.12a).

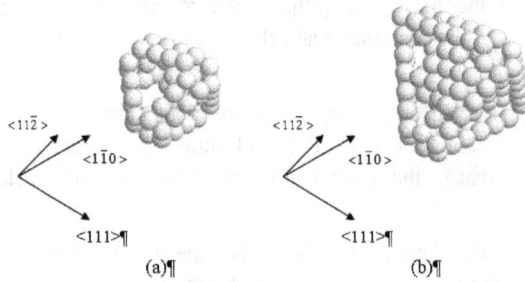

Figure 3.12: The tetrahedra of the packing defects, into which the spherical nanopores consisting of 15 (a) and 38 (b) vacancies are reconstructed, as a result of thermal activation.

In the case when the pore is an ideal tetradecahedron, it was possible to construct a geometrically correct tetrahedron of packing defects, even though the number of vacancies exceeded the required number. For example, the pore in Fig. 3.11a was rebuilt into a tetrahedron, but inside it, the lattice did not remain ideal but filled with vacancies (see Fig. 3.12b).

The number of vacancies that make it possible to build an ideal tetrahedron (..., 10, 15, 21, ..., and so on) is not sufficient for constructing an ideal tetradecahedron. Due to this, large pores cannot be reconstructed into a regular tetrahedron due to the fact that some vacancies break away from the pores as a result of thermal vibrations of the atoms. As a rule, symmetry breaking consisted in not completely building one of the edges or in the absence of the ideality of one of the vertices. Sometimes the pore could rearrange into two tetrahedra, having a common vertex. An important point is that when the computational cell was heated above the temperature $\approx 0.45 \cdot Tm$, all the spherical pores started to rearrange into the tetrahedron of packing defects until their radius began to exceed 13 Å (236 vacancies). A larger pore radius led to its collapse and the formation of dislocation loops.

Let us study the influence of atomic displacement waves on the processes of nanostructure reconstruction. To begin with, we generate the elastic waves in the computational cell which propagate with the velocity of longitudinal sound waves c_p (for gold $c_p = 3240$ m/s [91]). When creating a wave by assigning the appropriate velocity to the boundary atoms, the temperature of the computational cell rises. Therefore, the excess energy had to be removed with a thermostat. As a result, the temperature of the computational cell with some oscillations remained equal to the specified starting value. So, Fig. 3.13 shows the graph of the temperature change of the computational cell when the elastic waves are generated every 2.5 ps in the computer experiment.

If the elastic waves were generated in a computational cell containing a nanopore, then its transformation into a tetrahedron of packing defects was beginning. During the experiment, the initial temperature of the computational cell was set equal to 300 K. At this temperature, as already mentioned above, the spherical shape of the pore is metastable. The heating of the cells as a result of the formation of waves, as follows from Fig. 3.13, is insufficient to thermoactivate the process of pore reconstruction. Consequently, the stress caused by the elastic wave causes the transformation.

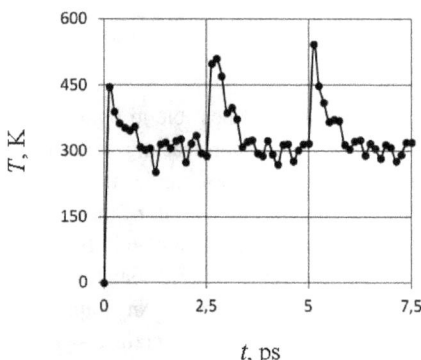

Figure 3.13: *The change in the temperature T of the computational cell in the experiment during the generation of sound waves at times 0, 2.5, and 5 ps. A proportional thermostat is used. The initial temperature of the computational cell is 300 K.*

One of the mechanisms for reconstructing the pores into the tetrahedron of packing defects is as follows. As a result of absorbing vacancies and reaching a certain critical radius, the nanopores can collapse in the most densely packed plane of the crystal, forming a dislocation loop. In the case of a low energy of the packing defect, the loop is transformed into a tetrahedron of the packing defect [117]. Note that for gold, the energy of the packing defect is lower than for metals such as silver, copper or nickel [118]. In our case, the restructuring begins because the sound wave causes some atoms, that form the surface of the pore, move to the nearest vacant sites, violating the initial form. Such a non-equilibrium state causes the pore to be restructured into an energetically more advantageous configuration, which is the tetrahedron of packing defects. The process of reorganization of the pore consisting of 38 vacancies (see Fig. 3.11a), which are under the influence of sound waves, is shown in Fig. 3.14 with the help of a visualizer of superposition of close-packed rows. In this case, for the transformation of the pore, there was enough of two waves generated in the computational cell at 2.5 ps interval. As follows from the analysis of Fig. 3.14b at the moment of crossing the pore by a wave, the vacancies are divided into two groups, each of which subsequently forms fragments of packing tetrahedra (see Fig. 3.14c). During the subsequent relaxation of the structure, a single and a larger tetrahedron is formed (see Fig. 3.14d). The subsequent generation of sound waves does not affect the resulting defect.

It should also be mentioned that when constructing the images in Fig. 3.14, the cooling procedure of the computational cell was used, by repeatedly zeroing the velocities of the atoms to improve the visualization of the structure, so the front of the transmitted wave is not noticeable.

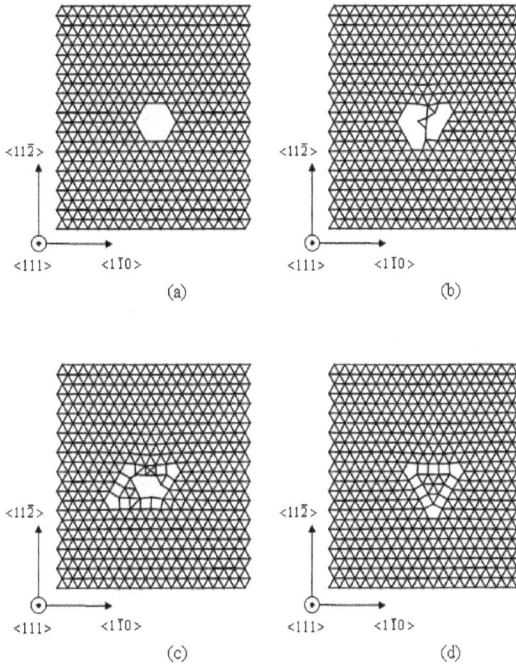

Figure 3.14: *The process of transforming the pores of their 38 vacancies into a tetrahedron of packing defects under the influence of sound waves. The fragments of the computational cell at the initial instant of time (a) and in 0.5 ps (b), 2 ps (c) and 5 ps (d) of the experiment.*

In computer experiments conducted at higher initial temperatures of the computational cell, the tetrahedra formation was also observed, but in some cases, as, for example, at 900 K, only one wave was sufficient. Obviously, in this case, the main contribution to the restructuring of the pore is made by the thermal vibrations of the atoms.

In order that the pore consisting of 116 vacancies (see Fig. 3.11b) be transformed into a tetrahedron of packing defects at a computational cell temperature of 300 K, it is required to pass about 20 sound waves through the crystal matrix. The increase in temperature reduced the number of waves necessary for the transformation.

Let us now consider the computational cells containing pores with a large radius. Thus, the pore consisting of 337 vacancies remained stable at a computational cell temperature of 300 K, regardless of the number of generated sound waves. The only thing that was observed in this case is the migration of atoms over the pore surface. This mechanism, apparently, is similar to the migration of atoms on the surface of a crystal, when for one elementary act the atom makes a movement many times greater than the interatomic one. Such a process was called the "roll-field" [119]. When the starting temperature of the computational cell is increased to 600 K, the waves succeed in splitting off a small group of vacancies from the "parent" pore (that is, created before the experiment), but later this group begins to transform into a double tetrahedron of packing defects (see Fig. 3.15).

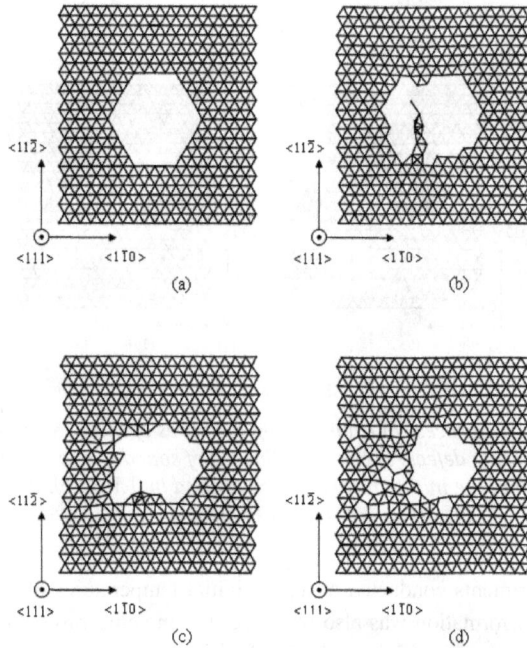

Figure 3.15: *Structural changes in the pore, consisting of 337 vacancies, after the passing of sound waves. The fragments of the computational cell at the initial instant of time (a) and in 0.5 ps (b), 2 ps (c) and 5 ps (d) of the experiment.*

At the next stage of the study, the shock waves were generated in the computational cell. To create such a wave, the outer atoms of the computational cell were assigned an initial velocity twice as large as the velocity of the sound waves. The study showed that at an initial temperature of the computational cell of 300 K, the shock wave initiates a pore reconstruction consisting of 38 vacancies into a double tetrahedron of packing defects. And part of the vacancies is split off from the pore, forming a small tetrahedron, and in the place of the pore, a tetrahedron of a larger size is formed. This splitting of vacancies is initiated by a shock wave, so a small tetrahedron is located in front of a large tetrahedron along the wave propagation path. Using the atomic displacement visualizer, which represents the lines connecting the initial and final positions of atoms, the changes in atomic positions are shown in Fig. 3.16 as a result of the structural transformation of the pore caused by the generation of ten sound (see Fig. 3.16a) and shock (see Fig. 3.16b) waves in the computational cell. A comparative analysis shows a violation of the symmetry of atomic displacements, caused by the separation of vacancies from the pore.

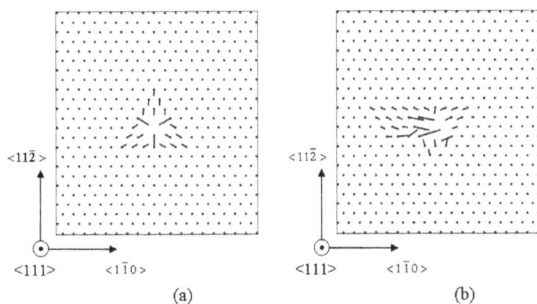

(a)   (b)

Figure 3.16: Atomic displacements near the pore consisting of 38 vacancies, after passing through the computational cell of ten sound (a) and shock (b) waves. The temperature of the cell is 300 K.

When the computational cell is heated to 600 K, the generated shock waves cause not only the transformation of the pore but also its subsequent displacement toward the wave source as a result of the relaxation of the stresses caused by the wave (see Fig. 3.17).

Note that the defect being transported consists in the fragments of a tetrahedron of packing defects. In the case of the alignment of a complete tetrahedron, apparently, the stresses formed would already be insufficient for its migration. It should also be noted that, as follows from the analysis of Figs. 3.16 and 3.17, the pore transformation and migration is a cooperative atomic process in which a large number of atoms are involved.

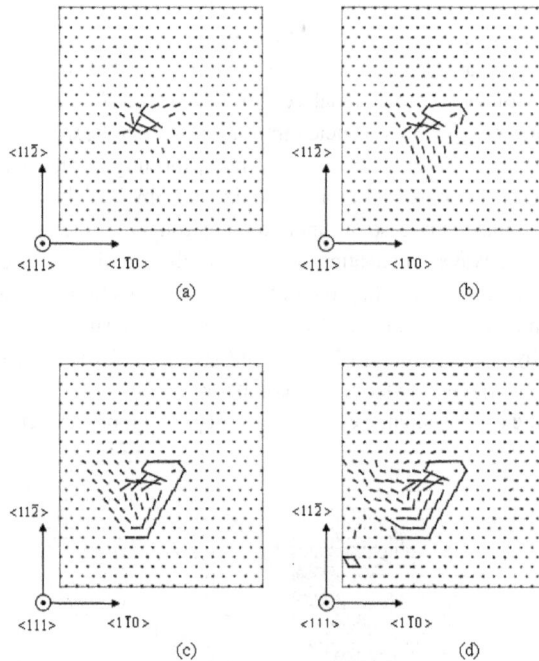

Figure 3.17: *Atomic displacements in the computational cell containing the pore of 38 vacancies after passing two (a), four (b), six (in) and eight (d) shock waves. The temperature of the cell is 600 K.*

When a shock wave passes through a pore consisting of 337 vacancies, the extreme atoms forming the pore surface can jump to remote empty nodes. As a result, about a third of the vacancies are split off from the pore and begin to move toward the source of the waves. Moving of the vacancies is carried out by throwing the nearest neighbouring atoms by the wave to vacant places. In this case, if the temperature of the computational cell is 300 K, the "parent" pore remains stable and motionless (see Fig. 3.18), and in the case of 600 K it begins to collapse (see Fig. 3.19). The destruction of the pore at an elevated temperature of the computational cell occurs not only because the shock waves split off groups of vacancies and thereby begin to break up the pore, but also due to the "evaporation" of vacancies by the pore. The process of "evaporation" appears to be simplified if the sphericity of the pore shape is disturbed since in this case, the surface energy increases. In addition, it should be noted that at a computational cell temperature of 600 K, a larger number of vacancies initially splits from the "parent" pore after the passage of waves than at 300 K, and a part of the split vacancies can form fragments of packing tetrahedra.

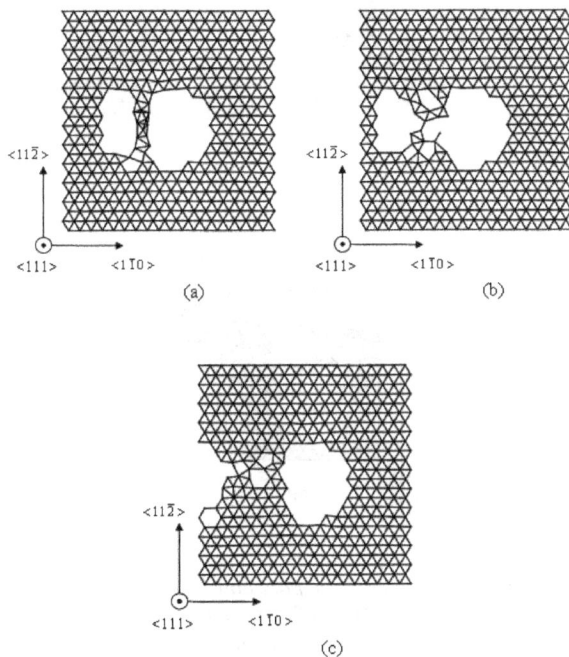

(a)   (b)

(c)

Figure 3.18: Structural changes in the pore consisting of 337 vacancies, after the passing through the computational cell of two (a), four (b) and six (c) shock waves. The temperature of the cell is 300 K.

Thus, the conducted research has shown that the passing through the crystal lattice of waves affects the structural changes of the nanopore. Thus, sound waves can cause the pores to be reconstructed into packing defect tetrahedra even at temperatures that are insufficient for an arbitrary pore rearrangement during the relaxation of the structure. The shock waves can cause the pore displacements, as well as their splitting or breaking. The processes of structural transformation of pores are affected by the number of vacancies, as well as the temperature at which these processes occur.

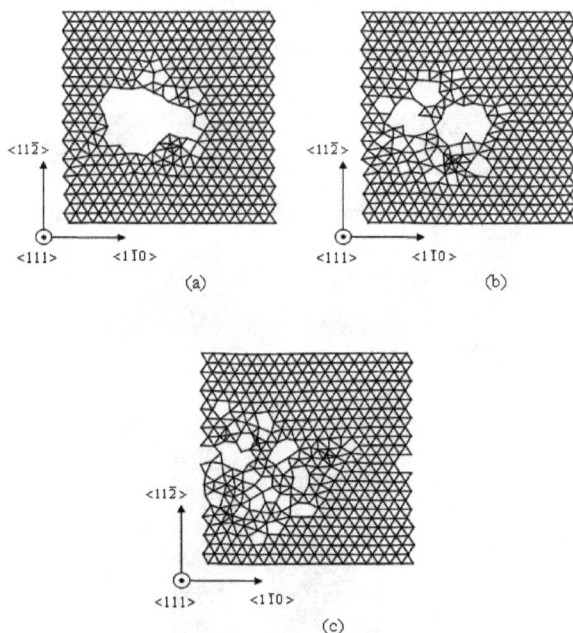

Figure 3.19: Structural changes in the pore consisting of 337 vacancies, after the passing through the computational cell of two (a), three (b) and four (c) shock waves. The temperature of the computational cell is 600 K.

3.4   Enlargement of nanopores

The basic mechanism of relaxation of metastable ensembles of radiation defects is nucleation and clustering. The processes of clustering must be taken into account when predicting the behaviour of structural units operating under various extreme conditions (high temperature, radiation exposure, shock loads, sudden temperature changes, etc.). In addition, the development of technologies that allow managing clustering processes is extremely useful in the synthesis of new materials with predetermined properties. As previously mentioned, the final stage of the evolution of the ensemble of vacancy pores is coalescence, which occurs as a result of the action of heat or radiation-induced mechanisms. The first mechanism is realized at elevated crystal temperatures, and the second at low temperatures. In this regard, the purpose of computer experiments conducted in this section was to study the effect of high-speed cooperative atomic displacements, considered as shock waves and sound waves, on processes of nanopore enlargement.

First, consider the process of merging two nanopores of equal size. To do this, in the computational cell that models the crystalline gold, we shall create two spherical nanopores consisting of 38 vacancies, so that they have a common edge. Recall that, despite the use of the term "spherical", the pores in the fcc lattice are of the tetradecahedron form. When performing computer experiments, the pores were arranged in such a way that the mentally held straight line connecting the centres of the tetradecahedrons coincided either with the crystallographic direction $<1\bar{1}0>$ (the "horizontal" arrangement of the pores) (see Fig. 3.20a) or with the direction $<11\bar{2}>$ ("vertical" arrangement) (see Fig. 3.20b).

As it turned out, the pore configuration data remain stable, up to a temperature of $\approx$ $0.45 \cdot Tm$. At higher temperatures, they are reconstructed into a single complex, which is an unfinished tetrahedron of packing defects (see Fig. 3.21). The violation of symmetry of the tetrahedron in our case consisted, as a rule, in the incomplete construction of an edge or in the absence of the ideality of one of the vertices.

Let us investigate the effect of atomic displacement waves generated in the computational cell on the coalescence of pores. In the beginning, let us consider the waves created by imparting the velocity, equal to the velocity of longitudinal sound waves c_p, to the boundary atoms of the computational cell. In order to exclude the thermal activation of the processes of structural pore reconstruction, we will set the computational cell a temperature equal to 300 K.

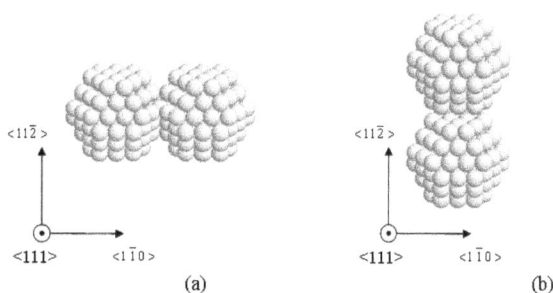

Figure 3.20: *Double vacancy pores located in the computational cell horizontally (a) and vertically (b).*

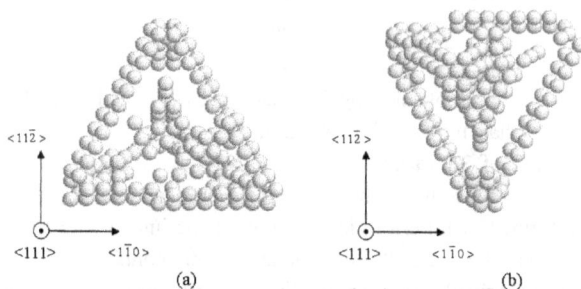

Figure 3.21: *The tetrahedra of the packing defects formed after 25 ps of structural relaxation of the computational cell at a temperature of 900 K (a) (pores were located horizontally) and 600 K (b) (pores were located vertically).*

The conducted research has shown that the length of the time intervals through which the waves are generated in the computational cell is of great importance for the processes of structural rearrangement. To simplify the computer experiments in each case, we will generate waves through equal time intervals. As the analysis of the calculations showed, in order to induce structural rearrangements of nanopores at temperatures insufficient for the thermal activation of diffusion processes, the time interval for generating waves must be relatively small. Thus, Figs. 3.22 and 3.23 show the structural changes in the pores located as shown in Fig. 3.20, after the passing through them of three waves generated in the computational cell every 2.5 and 10 ps of the computer experiment.

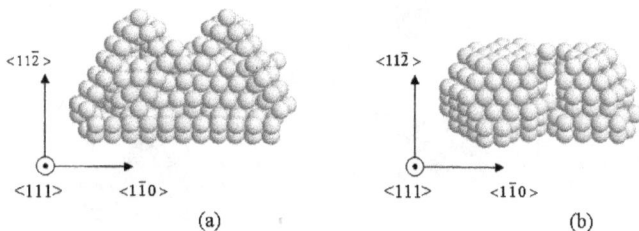

Figure 3.22: *Structural changes in the pores located horizontally, after the passing through the computational cell of three sound waves generated every 2.5 (a) and 10 (b) ps of the computer experiment.*

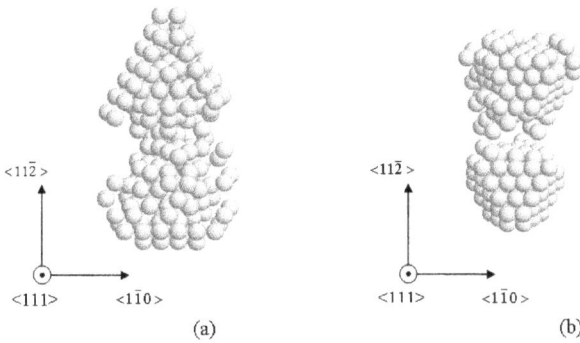

Figure 3.23: Structural changes of the pores located vertically, after the passing through the computational cell of three sound waves generated every 2.5 (a) and 10 (b) ps of the computer experiment.

As can be seen from Figs. 3.22 and 3.23, if the waves are generated every 2.5 ps in the experiment, then the structural transformations of pores take place. Thus, in the case of the initially horizontal arrangement of the pores, two non-ideal tetrahedra of packing defects are formed, which have one common edge (see Fig. 3.22a), and in the case of a vertical arrangement, only one of the pores is reconstructed into a tetrahedron (see Fig. 3.23a). These structural transformations are caused by the asymmetric displacement of atoms, which form the pore surface, from their equilibrium positions. As a result of the increase in surface energy, the vacancy system is removed from the metastable state, and assumes an energetically more feasible configuration. At longer intervals of generation, by the time the next wave approaches the pore, the atoms which are removed from the equilibrium positions by the previous wave, manage to mostly return to their original positions and, therefore, the structural pore transformations are practically not observed (see Figs. 3.22b and 3.23b).

It is obvious that a larger number of waves, in the long run, will cause a complete reorganization of the pores into a single complex. So, in Fig. 3.24, the structural changes in nanopores are presented after the passing of ten waves generated every 2.5 ps of in the computer experiment.

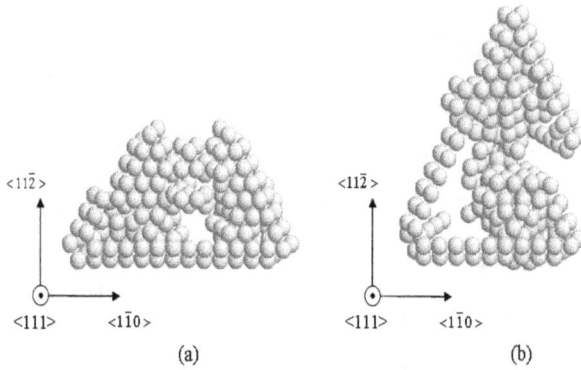

Figure 3.24: *Structural changes in the pores located horizontally (a) and vertically (b), after the passing through the computational cell of ten sound waves generated every 2.5 ps of the computer experiment.*

As can be seen from Fig. 3.24b, with the vertical arrangement of the pores, the passing waves cause their rearrangement into a single complex, which is a tetrahedron of packing defects, and the structure of the tetrahedron is more symmetrical on the side of the wave source. The vacancy complex presented in Fig. 3.24b is not an ideal tetrahedron, although some of its individual fragments are visible. The reason for this may be the fact that, with the horizontal arrangement of the pores, one of them shields the second one from the wave. Therefore, the impact on such a structure is less significant.

Let us turn to the study of the processes of coalescence of pores under the influence of shock waves created by imparting the velocity, which is twice the speed of sound waves, to the boundary atoms of the computational cell. Computer experiments have shown that the impact of such waves on the pores does not lead to the formation of a single complex. Thus, with a horizontal configuration, a pore rearrangement closer to the source of waves is observed, into an almost ideal tetrahedron of packing defects, and the second pore does not undergo significant structural changes (see Fig. 3.25a). In the case of a vertical configuration, both pores experience the influence of the wave in equal measure, as a result of which two separate tetrahedra touching the vertices are formed (see Fig. 3.25b).

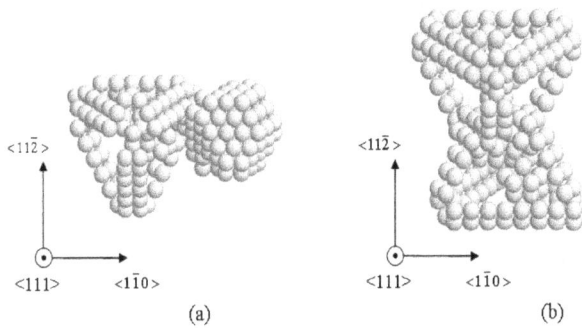

Figure 3.25: Structural changes in the pores located horizontally (a) and vertically (b), after the passing through the computational cell of three shock waves generated every 10 ps of the computer experiment.

We note that structural changes are observed at relatively large time intervals between the wave generations. This is due to the fact that the shock waves lead to a partial destruction of pores, and in the event that a significant number of steps of the computer experiment take place between individual waves, only then vacancies manage to form tetrahedra that remain stable under the impact of shock waves.

In the computer experiments described above, nanopores of equal size were considered. Obviously, the probability of a situation where two identical pores are nearby is not high. Nevertheless, this simplification allows us to obtain the necessary qualitative conclusions. Much more realistic is the situation when a separate vacancy or a small vacancy complex appears near the pores. As a result of the absorption of such defects, the pore size increases. As one of the mechanisms of pore growth under the influence of shock waves, the process of absorption of individual vacancies displaced by waves by small vacancy clusters can be proposed. According to the study, the small vacancy clusters, which are a double tetrahedron of packing defects, are the most stable configuration of the vacancies when the shock waves pass through the crystal structure, and the waves overcome them with minimal energy losses. The larger the vacancy cluster, the greater obstacle it becomes for the wave. As shown in the previous paragraph, when the waves encounter large pores, a part of vacancies is split off from the "parent" pore, which afterwards are carried away by subsequent waves. Thus, in the event that a small vacancy cluster is located near the pore, it will combine with the vacancies that have been split off. Let us verify this assumption experimentally. Near the nanopore, which consists of 236 vacancies, we arrange a complex of four vacancies (see Fig. 3.26a). The vacancies will be located on the front side of the pore that is, from the side facing the source of the waves. After the passing through the pore of three shock waves (waves were generated every 2.5 ps), partial capture of a small vacancy cluster is sometimes observed (see Fig. 3.26b). During the subsequent structural relaxation, the pore finally absorbs vacancies. For example, in the experiment described, the pore absorbs three vacancies out of four (see Fig. 3.26c). Consequently, we can assume that if the wave sources are located in the crystal arbitrarily, then, in the course of time, all the small vacancy clusters must be absorbed by the pore.

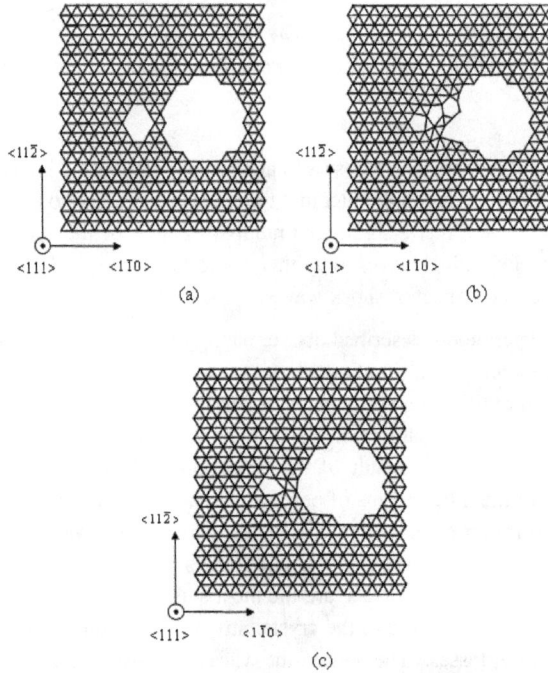

Figure 3.26: The process of absorbing by the pore of a small vacancy cluster under the influence of shock waves. The fragments of the computational cell at the beginning of the experiment (a) are presented after the passing of three shock waves generated every 2.5 ps of the experiment (b) and after 25 ps of structural relaxation (c).

Thus, the conducted study has shown the possibility of initiating processes of nanopore enlargement by the waves at temperatures not sufficient for the beginning of so-called heat-induced coalescence.

3.5   Dissolution of a nanopore near a free surface

Irradiation of various materials with ionic and plasma fluxes is actively used to modify and improve the properties of the surface layer, thus causing various changes in it: phase, structural, physical, mechanical, chemical and other. Depending on the parameters of the radiation flux and surface conditions, these changes can lead to the development or smoothing of the relief. Modification of the surface can be achieved either by direct ion or plasma action, or by the phenomena initiated by them, such as ion-induced stresses, dislocation mobility, recrystallization, changes in the composition of near-surface layers, and so on [120].

The defects of the crystal lattice formed at such high-energy impact can lead to a number of undesirable phenomena, one of which is the swelling of the material. The formation of a pore is a first-order phase transition in a point defect system since a new pore surface is formed in this case [121]. The driving force for the further diffusion evolution of vacancy pores is the desire to reduce the free surface. Two tendencies are distinguished: the coalescence of pores with a decrease in their total surface with an unchanged volume, when the pycnometric density remains constant ("internal" sintering), and the healing of individual pores with increasing pycnometric density ("external" sintering) [38]. The second tendency is especially pronounced when the pore is located near the surface of the crystal. In this case, during prolonged annealing at a constant temperature, a vacancy "evaporation" of the pore is observed.

It should be stipulated that the crystallites of a polycrystalline material, some of whose surface borders on the external medium, have different properties, in comparison with internal crystallites. These differences are caused by an increase in the specific surface energy, as well as by the processes caused by surface migration of atoms and their interaction with the intrinsic and foreign gas phases [22].

The purpose of computer experiments conducted in this section is to study the effect of shock and sound waves on the processes of structural changes in nanopores near the surface.

In the calculations, the same model as in the previous sections was used, but in which the free boundary conditions were set along one of the directions of the computational cell to create a free surface of the crystal, and periodic ones along the others.

In the study of a solid with a free surface, the main characteristic is the surface energy, which is important for determining the adhesion energy [122]. Let us define the surface energy for our model as an excess of potential energy when using the free boundary conditions per surface area. Calculations will be carried out under the free boundary conditions along the crystallographic direction <111>. It is obvious that the {111} plane has the lowest surface energy in comparison with other crystallographic planes, that is why the octahedral gold nanoparticles, as a rule, are faceted precisely by these planes [123]. The results of the calculations are given in Table 3.3.

The discrepancy between the calculated values and those already available can be explained by the fact that in literature sources the surface energy was calculated for macrocrystals, and the magnitude of this characteristic decreases with decreasing crystal dimensions [125]. In this regard, the results obtained can be considered satisfactory, and, consequently, the model used is completely applicable for further experiments.

Table 3.3. The calculated value of the surface energy σ for gold

T, K	σ, mJ/m^2	
	Obtained results	Literary data
1243	1192	1450 ± 80 [91]
1250	1198	1450 ± 20 [124]
1270	1211	1354 ± 50 [124]

Let us consider the process of dissolution of the pores near a free surface. It is known that it can flow through two mechanisms: vacancy dissolution of the pore and diffusion flow. The first mechanism is realized if the pore size is much less than the distance to the vacancy, and the second is if the pore size exceeds this distance [38]. In our experiments, the structural changes in the pores began at temperatures $\approx 0.45 \cdot T_m$. As a rule, the dissolution process was a combination of both of the above mechanisms. So, Fig. 3.27a shows a fragment of the computational cell containing the pore of 236 vacancies. After 25 ps of the computer experiment at a computational cell temperature of 900 K and when using the free boundary conditions in the direction <11$\bar{2}$>, some vacancies "evaporated". Moreover, along with the diffusion flow, which is a cooperative process of atomic displacements, the visualizer of atomic displacements also displays individual atomic jumps from the surface layer to the inner surface of the pore, which corresponds to the vacancy dissolution (see Fig. 3.27b). However, this mechanism is not dominant in our case.

We also note that when the centre of the nanopore is removed by one interatomic distance from the outer layer of atoms as well as the number of vacancies is decreased to

38, the effect of the surface practically ceases, since in this case the pore is transformed into a tetrahedron of packing defects.

Interest is caused by the change in the potential energy of the atoms of the surface layer during the dissolution of the pore. In order to present this change visually, we will draw maps of the level lines. When they are constructed, the value of the potential energy for each atom of the surface layer is determined, and then, depending on the arrangement of the atoms, a matrix of values is compiled. Further on, based on the elements of this matrix, linear interpolation of values is carried out, and lines of the same level are formed.

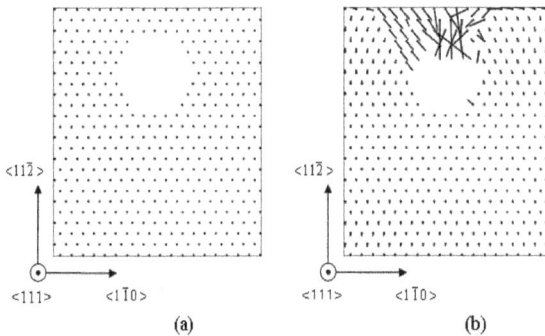

Figure 3.27: A fragment of the computational cell containing the pore of 236 vacancies at the beginning of the experiment (a) and after 25 ps of the computer experiment at a computational cell temperature of 900 K (b). Free boundary conditions are used in the crystallographic direction $<11\bar{2}>$.

Let's do the following experiment. In the computational cell, we create a nanopore consisting of 236 vacancies, so that its centre is located at six interatomic distances from the surface. The surface is the {111} plane. After a certain number of steps in a computer experiment performed at a temperature of 900 K, the value of the potential energy of the atoms of the surface layer is calculated and a level map is constructed, as shown in Fig. 3.28.

It should be stipulated that, for the sake of clarity, when plotting the maps in Fig. 3.28, fill areas were used instead of lines, in so doing a lighter section corresponds to a higher value of potential energy. The non-painted portion in the centre of the map is due to the absence of atoms in this layer, since the vacancies at this point escape to the surface of the crystal. For comparison, in the model used, the average potential energy per atom in an ideal computational cell is -3.91 eV.

Let us turn to the study of the effect of high-speed cooperative atomic displacements on the process of pore dissolution. The study showed that when the velocity of longitudinal sound waves c_p is imparted to the boundary atoms, then the dissolution of the pore does not occur. In separate experiments, when the pore was located at a certain distance from the surface, its rearrangement into a tetrahedron of packing defects was observed.

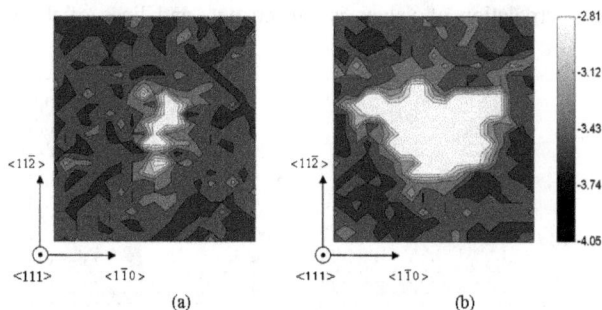

Figure 3.28: Maps of the distribution of potential energy (eV) of atoms of the surface layer in 10 ps (a) and 25 ps (b) of the computer experiment.

This process was initiated by the generated waves since the temperature of the computational cell was not sufficient for its thermal activation. In the event when the velocity of $2 \cdot c_p$ was assigned to the boundary atoms in order to create the wave, the results of the experiment turned out to be different. As a result of the passing through the computational cell of the generated shock waves, there was a splitting of the vacancy group from the "parent" pore and their subsequent escape to the free surface. Under the influence of such waves, the pore partially dissolved even at relatively low temperatures, for example, 300 K. An additional study, whose purpose was to determine the time intervals for generating shock waves, showed that the best result was achieved when creating waves through a large number of steps in a computer experiment, for example, 10 ps. Thus, Fig. 3.29 shows the result of the partial dissolution of the pore, consisting of 236 vacancies, by the shock waves (see Fig. 3.27a). The use of visualizers for the imposition of close-packed rows (Figs. 3.29a, b) and atomic displacements (Figs. 3.29c, d) allows us to draw the following conclusion. The splitting of the vacancy group from the "parent" pore by the shock waves initiates the cooperative atomic displacements that carry individual vacancies to a free surface. This process transforms the system into an energetically more advantageous state, since in the division of the pore an additional free surface is formed inside the computational cell, and its dissolution reduces the surface energy.

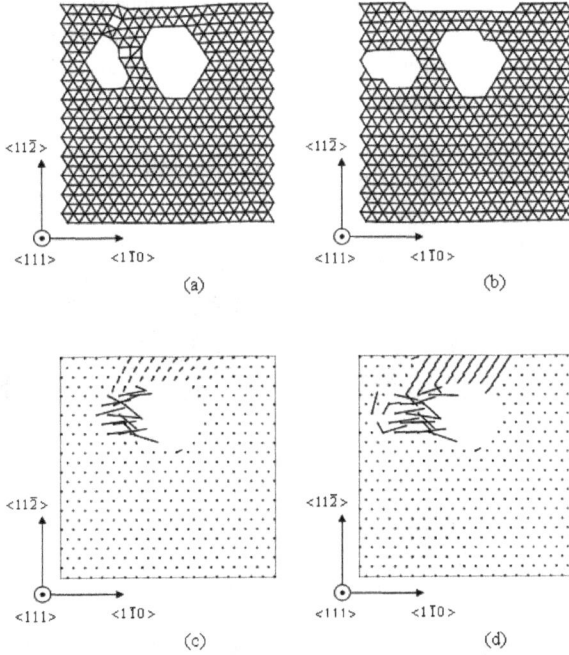

Figure 3.29: *The process of partial "evaporation" of the pore, consisting of 236 vacancies, near a free surface under the influence of shock waves. Fragments of the computational cells represented by various visualizers are presented after the passing through of three (a, c) and five (b, d) waves. The temperature of the computational cell is 300 K.*

We also note that, because of the use of the thermostat, the computational cell does not warm up when the waves are generated, and its temperature remains close to the set value. Consequently, the thermal activation of the pore dissolution process is absent. Therefore, there is no vacancy "evaporation" in Fig. 3.29 since the temperature of the computational cell is not sufficient to initiate this mechanism.

Fig. 3.30 shows the maps of the potential energy distribution of atoms for the experiment described above. Fig. 3.30a indicates the presence of a zone of increased energy that connects the "parent" time with the split off part of the vacancies. This zone is formed at the time of separation of the pore and subsequently disappears (see Fig. 3.30b). The greatest number of cooperative atomic displacements leading to the dissolution of pores occur precisely at the time when the pore is in an intermediate and unstable state, the state of the division into two components.

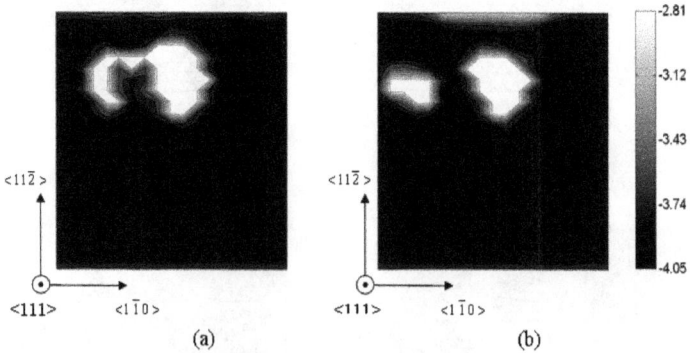

(a)   (b)

Figure 3.30: Maps of the distribution of potential energy (eV) of atoms in 30 (a) and 50 (b) ps of the experiment. These time points correspond to the computational cells shown in Fig. 3.29a and 3.29b.

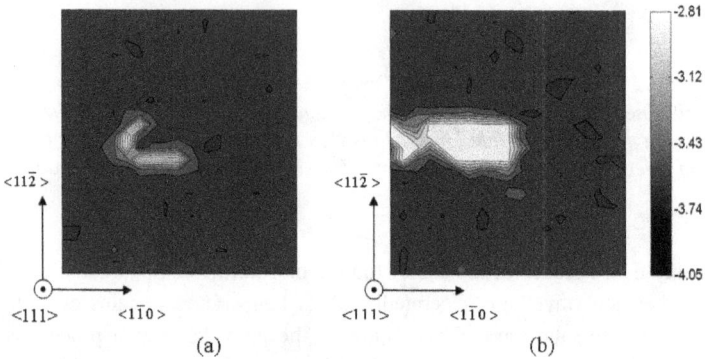

(a)   (b)

Figure 3.31: Maps of the distribution of potential energy (eV) of surface layer atoms in the computational cell containing a pore of 236 vacancies, after the passing of two (a) and five (b) shock waves. The temperature of the cell is 300 K.

An additional study showed that the process of pore dissolution under the influence of shock waves is observed, no matter what a crystallographic plane of the computational cell is chosen as the surface. For example, Fig. 3.31shows the maps of the potential energy levels of the atoms of the computational cell, after the passing through a pore consisting of 236 vacancies of two and five shock waves, using free boundary conditions in the <111> direction and an initial temperature of 300 K.

As follows from the analysis of Fig. 3.31, in the absence of vacancy evaporation, the "escape zone" of vacancies to the surface is a clearly delineated region of small size, oriented along the close-packed direction to the wave source. In this case, with the help of visualizers used in the construction of images in Fig. 3.29, it is not possible to obtain a clear picture of the described process, so we will limit ourselves only to the images of the potential distribution of the atomic energy.

Thus, the experiments described in this section indicate that the process of dissolution of the nanopore can be realized by generating shock waves in the crystal structure. In this case, dissolution is carried out even at temperatures insufficient to initiate thermal activation of the diffusion dissolution mechanism.

3.6   Peculiarities of structural transformations of a nanopore in a deformed computational cell

The details of various devices operating under conditions of intense radiation exposure are exploited under considerable loads, which create inhomogeneous stresses in the structural material. For example, it is known that TVEL envelope undergoes various power actions, including vibration loads, installation forces, the internal pressure of gaseous fission products, swelling fuel pressure, thermal stresses in the shell [126], and so on. In addition, with the accumulation of radiation defects, additional stresses in the material may appear due to the inhomogeneity of the swelling.

The computer experiments conducted in this section are devoted to determining the effect of shock waves on the structural transformations of nanopores in a deformed computational cell. Deformation was modelled by changing the equilibrium interatomic distances in the computational cell. Only elastic deformation was considered.

It is known that the increase in the free surface during a pore collapse is compensated by the decrease in its energy due to the restoration of bonds at the interplanar distance [118]. That is why the dislocation loop formed during the collapse is located in the plane of close packing. The subsequent cleavage of the loop leads to the formation of a tetrahedron of packing defects. Healing of pores under the influence of stresses is carried out by a dislocation mechanism. In this case, shear stresses arise near the pore, and in the case of exceeding the critical values of the Frank-Read source, a dislocation loop is formed and the pore boundary shifts by the Burgers vector value [127]. The computer experiments carried out showed that Frank's dislocation loop, despite the high energy barrier, begins to nucleate at a temperature of $\approx 0.45 \cdot T_m$, and its source is the internal surface of the pore. Since gold is a metal with a low energy of packing defects, splitting of loops is subsequently carried out, and tetrahedra of packing defects are formed (see Fig. 3.32). The generated dislocation loop should lower the surface energy of the pore to a value that allows it to remain stable at a given temperature. The size of the loop depends on this.

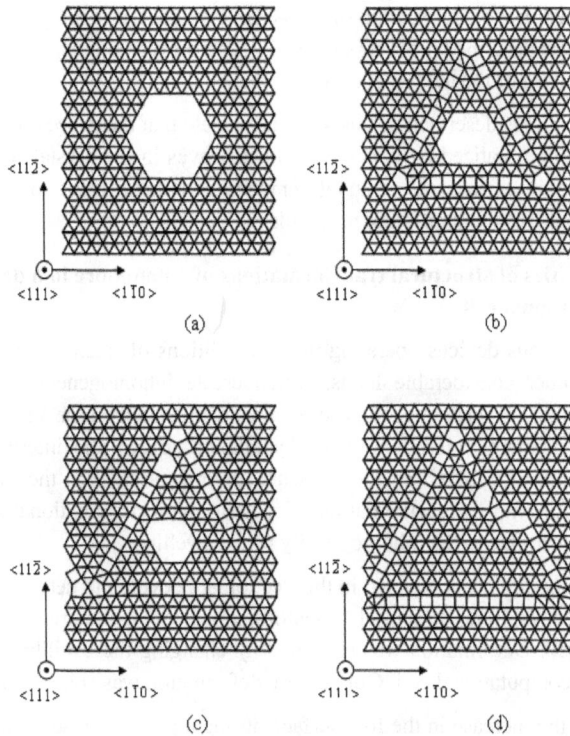

Figure 3.32: *Tetrahedra of packing defects, obtained as a result of the splitting of the Frank dislocation loop. A fragment of the starting configuration of the computational cell containing the pore of 236 vacancies (a) and also through 25 ps of the experiment at a temperature of 675 (a), 750 (b), and 875 (c) K.*

Computer Modelling of Structural Transformations of Nanopores in Fcc Met.   Materials Research Forum LLC
Materials Research Foundations **63** (2019)   https://doi.org/10.21741/9781644900512

We shall investigate the effect of uniaxial deformation of a crystal and its temperature on the structural transformations of nanopores. To do this, the computational cell containing the nanopores of different sizes was deformed along one of the axes and then held for 25 ps of the computer experiment at a given temperature. Later, the procedure of zeroing the velocities of atoms followed, which resulted in the system passing to a state with a local minimum of potential energy, and then visually studying the resulting configuration of atoms in order to reveal structural changes in the pore. Similarly, the minimum temperature T_m was determined, at which the pore was transformed into a tetrahedron of packing defects as a function of the uniaxial deformation ε_x (see Fig. 3.33). The approximation of the dependencies in Fig. 3.33 was performed using polynomials of the third degree.

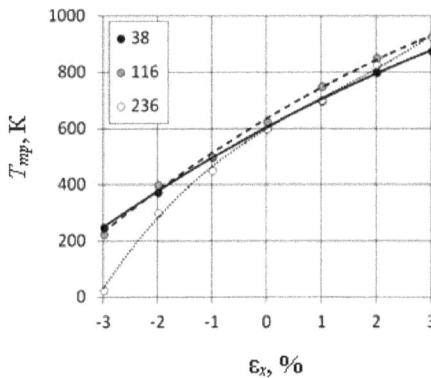

Figure 3.33: *Dependence of the temperature T_{mp}, at which the pore, consisting of a different number of vacancies, begins to transform into the tetrahedron of packing defects, on the uniaxial deformation ε_x.*

As follows from the analysis of the dependencies given in Fig. 3.33, the tensile stresses stabilize the pore, as a result of which the minimum temperature T_{mp}, at which the structural transformation of the pore begins, increases. In addition, the value of the surface energy of the pore consisting of 236 vacancies is sufficient to induce a rearrangement under the impact of compressive stresses practically without thermal activation.

It is obvious that the elastic properties of the crystal in different crystallographic directions will not be the same. Nevertheless, the conducted study showed that if the computational cell containing the pores with 38 or 116 vacancies deforms within the limits shown in Fig. 3.33, then the differences in the transformation temperature of the pore T_{mp} obtained under the uniaxial deformation along different directions, do not exceed 25 K, which is only 5% of the mean value of T_{mp}. Therefore, the dependence $T_{mp}(\varepsilon)$ is given only for deformation of the computational cell along the X-axis corresponding to the crystallographic direction $<1\bar{1}0>$.

It should be stipulated that in the case of deformation of a computational cell containing a pore of 236 vacancies along the Z -axis, the difference in the values of T_{mp} from the data obtained by deformation in the other two directions can reach 100 K. However, an additional study carried out using the regression analysis showed that at a significance level of α = 0.05, the differences between the T_{mp} dependences, in this case, can be considered statistically insignificant.

Structural transformation of the pore is a cooperative process of atomic displacements, which within the framework of the considered model is realized during several thousand steps of a computer experiment. Obviously, at the same temperature of the computational cell, the deformation will affect the duration of this process. Thus, Fig. 3.34 shows the time dependence of the time t_{mp} necessary for rearrangement of the pore in the tetrahedron of packing defects, on the value of the uniaxial deformation ε_x. The temperature of the computational cell was set equal to 900 K since, with this value, the transformations of all the investigated pores in the strain range under consideration are observed. Approximation of the obtained data was carried out using exponential functions.

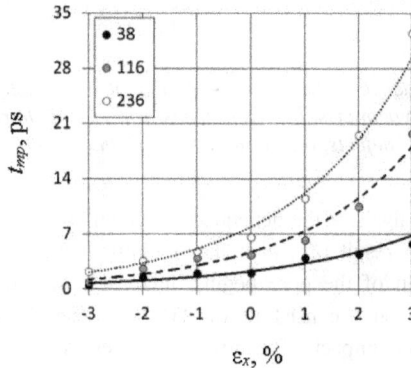

Figure 3.34: *Dependence of the duration of the transformation process t_{mp} of the pore, consisting of a different number of vacancies, on the magnitude of uniaxial deformation ε_x. The temperature of the computational cell is 900 K.*

According to the obtained dependences, which are shown in Fig. 3.34, the duration of the pore reconstruction process increases with the increase in its size, which is obviously due to a large number of displaced atoms. The compressive stresses not only initiate the accelerated transformation of the pore but also reduce the difference in the time of transformation of pores of different sizes. Tensile stresses, as mentioned earlier, maintain the shape of the pore, as a result of which it remains metastable for a long time interval, despite the fact that the temperature of the computational cell is high enough.

Note that in order to calculate the time of the pore transformation, the visual analysis used in the construction of the dependencies in Figure 3.33 is not enough, therefore, the change in the potential energy of the computational cell in the process of structural relaxation was calculated, and the sought-for moment was determined from local minima. Let us explain it with an example. For instance, Fig. 3.35 gives graphs of the variation in the computer experiment of the potential energy of a computational cell containing a pore of 38 vacancies, when it is heated to 900 K, and also for different values of uniaxial deformation.

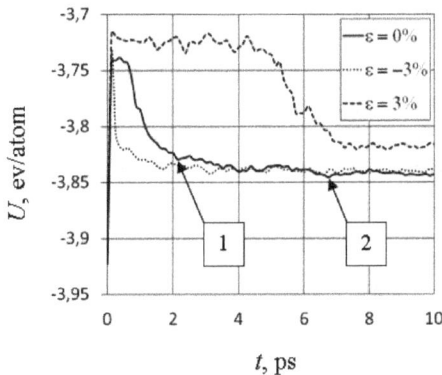

Figure 3.35: The change in the potential energy of the computational cell U during the experiment for a different value of the deformation along the X-axis. Local energy minima are indicated for the pore transformation.

As can be seen from Fig. 3.35, the process of pore transformation is accompanied by a sharp decrease in the potential energy of the computational cell. The first local minimum indicated in the figure corresponds to the formation of fragments of packing defects, which are the edges of the tetrahedron (see Fig. 3.36a). At the same time, an energy gain of 2.3% is achieved. The subsequent decrease in the potential energy is due to the ongoing process of aligning the tetrahedron, the arrangement of the atoms inside which begins to correspond to an ideal lattice (see Fig. 3.36b) by the time the second local energy minimum is reached, indicated in Fig. 3.35.

Let us now consider the effect of the shock waves on the process of structural transformation of a nanopore in a computational cell subjected to elastic deformation. Carried out computer experiments have shown that in the case of compression of the computational cell, the vacancies which are split off from the pore by the wave, form fragments of packing defects tetrahedra. In the case of the tension of the computational cell, a decrease in the number of vacancies that can be removed is observed, and with a subsequent increase in tensile stresses, the pore begins to stretch under the action of waves, but its integrity, as a rule, remains. Thus, Fig. 3.37 shows the results of the experiment in which the computational cell containing the pore of 236 vacancies was deformed along the X-axis, and five shock waves were generated with an interval of 5 ps at a temperature of 300 K. The presented images show that in the absence of deformation, the shock waves split off the vacancy group from the pore, as was already mentioned earlier (see Fig. 3.37a). In the presence of compressive stresses, the splittable vacancies form an imperfect tetrahedron of packing defects displaced to the source of waves, and the "parent" pore is located at one of its vertices (see Fig. 3.37b). When the cell is stretched, the pore stabilizes, as a result of which the shock waves could not split vacancies off it, however, a significant change in the shape of the pore is observed (see Fig. 3.37c).

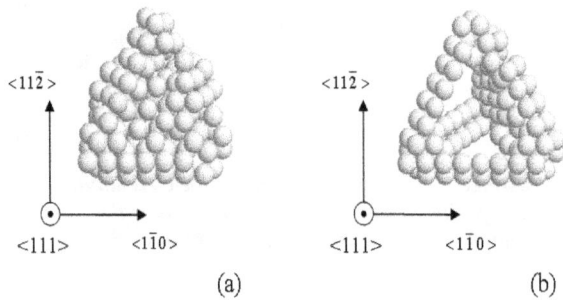

Figure 3.36: Structural transformations of the pore consisting of 38 vacancies during the computer experiment at a temperature of 900 K. The configuration of the atoms corresponds to the first (a) and second (b) a local minimum of the potential energy in Fig. 3.35.

If in the previous experiment the extension of the crystal led to the stretching of the pore due to the displacement of individual vacancies under the impact of the wave, in the case of pores of smaller dimensions, in some cases the movement of the entire cluster was fixed, without any significant destruction of the configuration. This fact is due not only to a decrease in the packing density of atoms under the action of tensile stresses but also by compensation of the surface energy of the pore by these stresses. In the described case, the relaxation of the stresses created by the wave is due to the displacement of the pore.

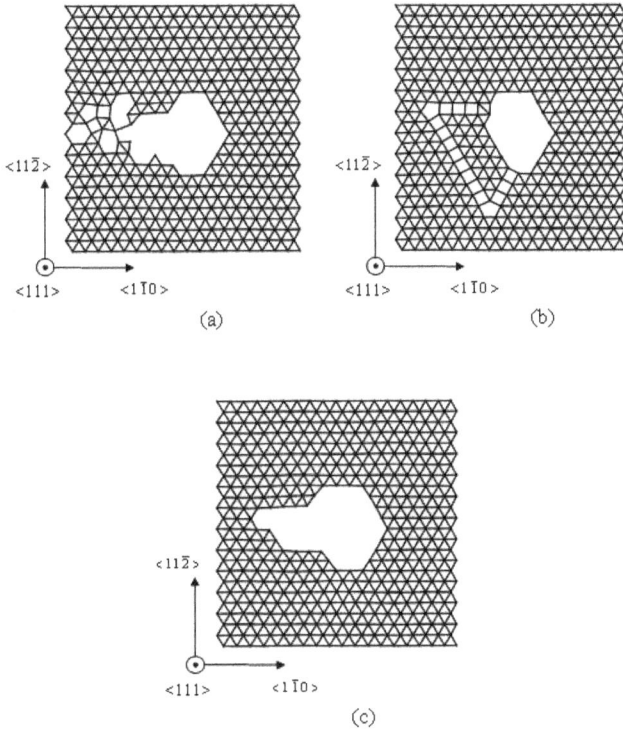

Figure 3.37: *Fragments of computational cells containing a pore of 236 vacancies, after the passing of five shock waves generated at an interval of 5 ps. The computational cells are subject to uniaxial deformation along the X-axis by the values ε_x = 0% (a), -1% (b) and 3% (c). The initial temperature was set at 300 K.*

Previous research has shown that, in addition to fragmentation of the pores into individual components, the shock waves can initiate a structural rearrangement of two pores into twin packing defect tetrahedra. It is obvious that the stresses created as a result of the deformation of the computational cell can contribute to the process of coalescence of the pores. To confirm the above, we will carry out the following experiment. In the computational cell subjected to uniaxial deformation, we create two closely spaced pores consisting of 38 vacancies, as shown in Fig. 3.20. After that, we generate the shock waves with an interval of 10 ps by specifying the initial temperature of the computational cell of 300 K. Structural changes of two nanopores, which are located in the computational cell along the crystallographic direction $<1\bar{1}0>$ ("horizontally") and along the direction $<11\bar{2}>$ ("vertically"), after the passing of three shock waves, are shown in Figs. 3.38 and 3.39, respectively.

It can be seen from the images obtained that in the computational cell subjected to compression, the process of pore fusion is observed with the formation of packing defect tetrahedra (see Figs. 3.38a and 3.39a). Similar results are obtained independently of the axis of compression of the computational cell. We note that the combination of pores in a single complex under deformation is observed only after the generation of shock waves since the temperature of the computational cell was not sufficient for the thermal activation of this process. Tensile stresses prevent the pores from merging, but under the impact of shock waves, they begin to shift either separately (see Fig. 3.38b), or together (see Fig. 3.39b). Particularly interesting is the first case. So, if we assume that the wave sources in the crystal are arranged in an arbitrary manner, then, under the influence of several waves, the pores can be gradually removed from each other, which will slow the coalescence process and, possibly, lead to a decrease in the swelling of the material.

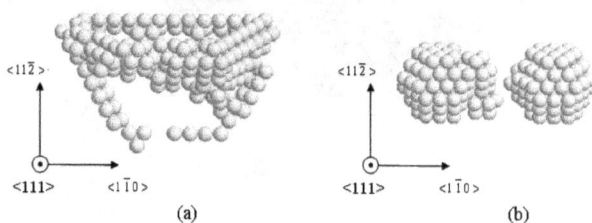

Figure 3.38: *Structural changes in the pores located horizontally after the passing of three shock waves generated at an interval of 10 ps, with the value of the deformation of the computational cell $\varepsilon_y = -2\%$ (a) and $\varepsilon_z = 3\%$ (b) of the computer experiment.*

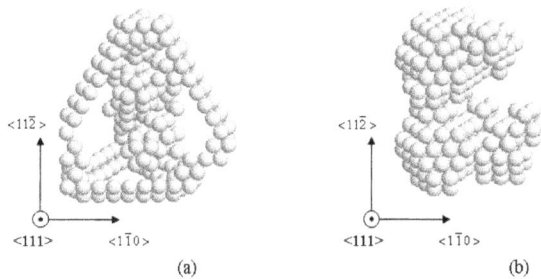

$<11\bar{2}>$ $<11\bar{2}>$

$<111>$ $<1\bar{1}0>$ $<111>$ $<1\bar{1}0>$

(a)   (b)

Figure 3.39: *Structural changes in the pores located vertically after the passing of three shock waves generated at an interval of 10 ps, with a value of the deformation of the computational cell,* $\varepsilon_y = -2\%$ *(a) and* $\varepsilon_z = 3\%$ *(b).*

Thus, the conducted research has shown that the impact of shock waves on nanopores in a deformed computational cell leads to new effects. With compressive stresses, the role of the surface energy of the pores increases, due to which the waves activate the processes of its structural transformations, which in some cases can lead to the fusion of individual pores. The tension of the computational cell stabilizes the pore, as a result of which the shock waves can cause its displacement without destruction.

3.7 Splitting of a nanopore in the grain boundary region

The materials from which the various structural units are made, as a rule, are polycrystals consisting of individual grains. Under the influence of irradiation in the grains, point defects and various trasmutants (e.g. of inert gases) are formed. At low temperatures, interstitial atoms, which have greater mobility than vacancies, migrate to the drains, which are the grain boundaries, and form impurity segregations that fix dislocations. As a result, the yield strength of the material rises. An increase in temperature leads to an increase in the diffusion mobility of vacancies, and the formation of porosity along the grain boundaries causes radiation embrittlement [17]. Thus, grain boundaries play an important role in the consideration of various radiation-stimulated phenomena.

The purpose of computer experiments conducted in this section was to determine the effect of post-cascade shock waves on the processes of structural changes in nanopores in a polycrystal.

In this work, a symmetrical tilt grain boundary is considered. To create the boundary, the computational cell was divided into two blocks, after which they were misoriented by an angle $\theta/2$ along the crystallographic direction <111>, convergence and removal of atoms located at a critically close distance. In order to minimize the energy of the boundary, it is possible to carry out a rigid shift of one grain as a whole with respect to the other in a direction parallel to the plane of the boundary, so that the boundary atoms find themselves at the positions of the local minimum of the potential energy. In the future, we shall consider both models of boundaries - without a shift and with a shift. Thus, for the first model, the "atomic" relaxation was used, in which each atom moves under the influence of all the forces acting on it until the sum of the energies of all pair interactions reaches a minimum, and for the second model, the relaxation was carried out in two stages: the "hard" relaxation, at which the sum of the interactions is minimized as a result of the grain shift, but each atom still occupies the initial node in its grain, and then the "atomic" relaxation follows [128].

When considering the system of two blocks, the misorientation of which leads to the formation of a grain boundary region, one of the main characteristics is the specific energy of the grain boundary γ, which is defined as the difference between the energies of a bicrystal and a single crystal containing the same number of atoms per unit area of the grain boundary. Specific energies of the grain boundaries, determined at different grain misorientation angles for the models used both without shear and with shear, are presented in Table 3.4.

Table 3.4 The specific energy γ of the tilt grain boundary for the considered models, depending on the misorientation angle θ, J/m^2

Misorientation angle θ	Boundary model without relative grain shear	The model of the boundary with relative grain shear
4°	0.539	0.461
6°	0.593	0.502
8°	0.596	0.536
10°	0.600	0.561

As can be seen from Table 3.4, the values of the specific energy γ for the models used increase with an increase in the misorientation angle θ. In addition, when using the two-stage relaxation ("hard" and "atomic"), one achieves a lower energy of the boundary.

The values of the specific energy of tilt grain boundaries for gold, determined experimentally, are equal to 0.364-0.406 J/m^2 [110], which is somewhat lower than the energy values determined as a result of computer simulation. Nevertheless, this difference is not critical and the models considered are fully applicable for further research.

First, let us consider a computational cell in which a spherical nanopore is located on the tilt boundary of two grains. The tilt grain boundaries are usually represented as the wall of edge dislocations. Obviously, in this case, the relative location of the pore and the grain boundary dislocation should influence the energy of the grain boundary. Thus, Fig. 3.40 shows a change in the specific energy γ of the grain boundary when the pore is displaced along the Y-axis by a distance ΔS from the edge of the computational cell.

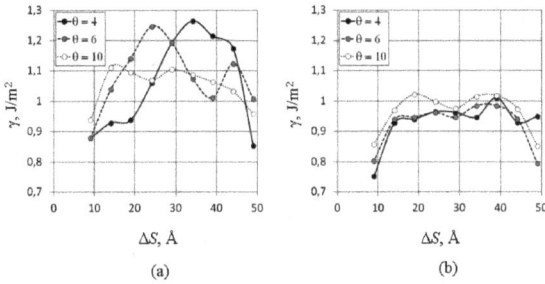

Figure 3.40: *Dependence of the specific energy γ of the tilt grain boundary on the distance ΔS between the centre of the pore and the edge of the computational cell along the Y axis for the model without grain shear (a) and a model with grain shear (b)*

It must be stipulated that sharp declines at the beginning and end of the curves are due to the fact that the pore is partially beyond the region for which energy is calculated.

As follows from the analysis of the dependencies constructed in Fig. 3.40, for the model with a relative grain shear the change in the location of the pore at the boundary practically does not affect the value of the energy. For the model in which the shear is not used, a significant effect of the location of the pore on the energy of the grain boundary is observed only for a small misorientation angle θ. The reason for this is the following. At a small angle of misorientation, the successive displacement of the nanopore along the boundary leads to the fact that it can take place between the nuclei of two grain boundary dislocations. This location of the pore corresponds to a local maximum on the curve (see Fig. 3.40a). The local minimum is observed if the centre of the pore is aligned with the dislocation core. When considering a larger misorientation angle, an increase in the density of grain-boundary dislocations led to the fact that the intersection of the pore and the dislocation core was always observed, so the dependence is more uniform (see Fig. 3.40b). In connection with this, in future investigations, we will consider only low-angle boundaries, since at high angles it is difficult to reveal the location of individual grain-boundary dislocations, and as a consequence, it will be difficult to determine the effect of an isolated dislocation on the pore.

We now turn to the study of the impact of the shock wave on the structural transformations of a pore located in a grain boundary region using a model without shear. The study showed that when pores are located between the nuclei of grain boundary dislocations, the shock waves succeed in splitting off a significant part of the vacancies from it (see Fig. 3.41). If, however, the centre of the pore coincides with the dislocation core, then for its splitting, higher-intensity waves are required, therefore, under the same experimental conditions, a smaller number of vacancies are split off in this case (see Fig. 3.42).

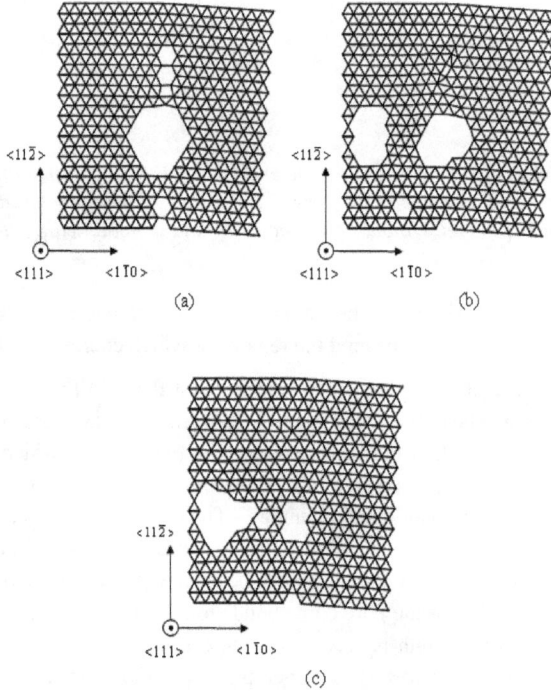

Figure 3.41: Fragments of computational cells with the pore located between the cores of grain boundary dislocations, at the beginning of the experiment (a), and after the passing of six shock waves obtained by assigning velocities equal to 1.6·c_p (b) and 2.8·c_p (c) to the boundary atoms. The temperature of the cell is 300 K.

It was said earlier that the location of the pores between grain-boundary dislocations is the least energetically feasible position. Therefore, in order to fragment the pores, the small stresses created by the wave are sufficient, and, consequently, when generating a wave in the computational cell, it is possible to assign atoms a velocity much lower than using a crystallite model. When the pores are combined with the dislocation core, it becomes more stable, and large stresses are required now for its splitting, therefore, structural changes are not observed in Fig. 3.42.

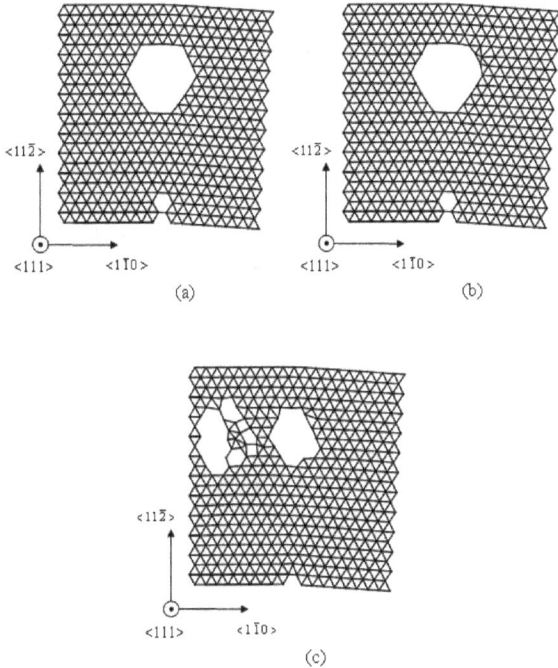

Figure 3.42: Fragments of computational cells with the pore, which is combined with the core of the grain boundary dislocation, at the beginning of the experiment (a), and after the passing of six shock waves obtained by assigning velocities equal to $1.6 \cdot c_p$ (b) and $2.8 \cdot c_p$ (c) to the boundary atoms. The temperature of the cell is 300 K.

We note that in the experiment described above a model of the tilt grain boundary with an angle of misorientation $\theta = 4$ was used. In the case of models with high angles, the earlier conclusions remained correct, but since the density of grain boundary dislocations increases, the pore in most experiments crossed their nuclei, which reduced the structural transformations caused by the waves. In addition, when carrying out the experiments with computational cells containing pores of smaller size, in some cases, their displacement from the grain boundary region was observed without destroying the structure of the pore itself.

If we carry out similar experiments using a model in which a relative grain shear is used, then the position of the pore in the grain boundary region will no longer matter (see Fig. 3.40b). Therefore, the pore splitting could be achieved by creating a wave with parameters, the magnitude of which almost coincides with those that are possessed by a wave that causes pore fragmentation in the crystallite model.

Let us now consider the computational cell in which the pore being created is located in one of the grains. An investigation of the stability of such a configuration showed that the pore remains stable and the diffusion drift of individual vacancies during the structural relaxation to the grain boundary was not observed. It was also noted in [129] that vacancies, as compared with interstitial atoms, interact relatively weakly with intergranular boundaries. The drift of vacancies, in this case, is observed when the computational cell is heated to a temperature close to the melting point of gold since, in this case, the pore begins to rapidly evaporate vacancies. Thus, the stresses created by the grain boundary are not sufficient for structural changes in the pore; therefore, when it is split into several parts by the shock waves, each component can already be regarded as a separate vacancy cluster, since at low temperatures they remain stable, lose contact with each other and merging into a single complex at the grain boundary does not occur.

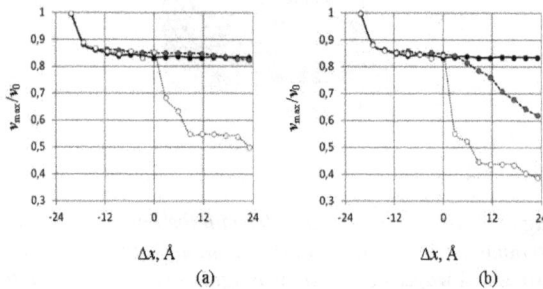

Figure 3.43: *The dependence of the speed fraction v_{max}/v_0 (v_{max} is the maximum velocity of the atom during the experiment, v_0 is the velocity assigned to the first atom of the chain equal to $2 \cdot c_p$) of the chain atoms on their coordinate Δx relative to the grain boundary with the misorientation angle θ equal to $4°$ (a) and $10°$(b). In the calculations we used the single crystal model (●), the boundary model without shear (●) and with the shear (○).*

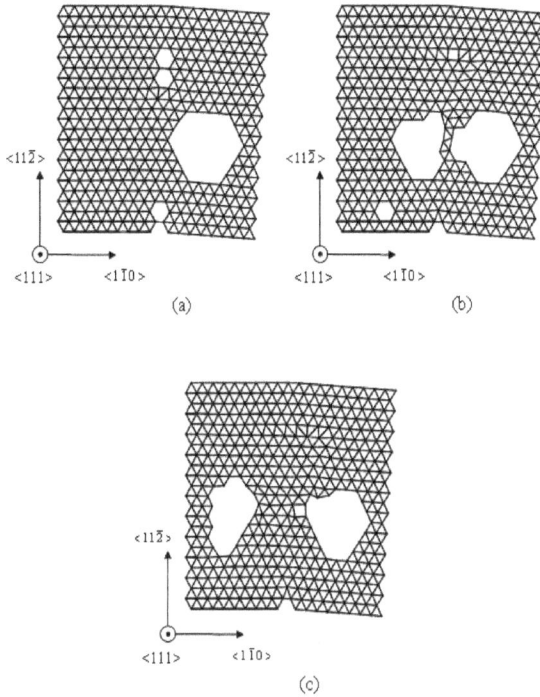

Figure 3.44: *Fragments of computational cells with the pores located in one of the grains at the beginning of the experiment (a), and after the passing of five (b) and nine (c) shock waves obtained by assigning to the boundary atoms a velocity equal to 2.8·cp. The temperature of the computational cell is 300 K.*

Let us study the possibility of the shock wave to cause structural transformations of the nanopore when separated from the wave source by the tilt grain boundary. Note that the boundary is not an insuperable obstacle to the waves. Therefore, first, we determine the loss of velocity of the wave of atomic displacements when crossing the boundaries obtained by using the models described above. To do this, in the computational cell, select a chain of atoms so that the boundary divides it in half. Assigning the initial atom a velocity v_0 in the direction $<1\bar{1}0>$, and thus creating relay-like atomic collisions, we determine the maximum value of the velocity v_{max} of each atom in a given direction. The results of calculating the velocity of atoms isolated in the computational cells with different grain misorientation are shown in Fig. 3.43. As follows from the dependences constructed, at a low angle of misorientation and the absence of a grain shear, the velocity of atoms in collisions practically does not differ from the rate of their collisions in a single-crystal lattice (see Fig. 3.43a). Therefore, using this model, the grain boundary will not be an obstacle to the wave. Using a model with a shift causes a decrease in the velocity (about 35%) of the atoms, which is caused by a violation of the focusing of atomic collisions during the transition from one grain to another. Another cause of defocusing may be the stresses created by the cores of grain boundary dislocations. An increase in the angle θ causes even a greater decrease in the velocity (see Fig. 3.43b). Nevertheless, with a low angle of misorientation of the grains and by imparting a high velocity to the atoms when creating the waves, it is possible to split a group of vacancies from the pores and transfer it through the grain boundary region (see Fig. 3.44).

It is known that the dissolution of the pore can be realized due to vacancy evaporation or due to the diffusion flow of material deep into the pore. The process of reducing the pore size under the influence of shock waves described in this research does not fit within the frameworks of these mechanisms. Therefore, we will call it dynamic dissolution, since it is initiated by high-speed cooperative atomic displacements.

Thus, the conducted research has shown that the post-cascade shock waves can cause the nanopore to be fragmented into separate components when it is located in the grain boundary region, while there are enough waves of much lower intensity compared to a single crystal. In many respects, the manifestation of this effect depends on the mutual arrangement of the pores and the grain-boundary dislocations. Also, at low misorientation angles, it is possible to have a through migration of the vacancy clusters through the boundary of the tilt grains.

3.8   Structural transformations of a nanopore of a cylindrical shape

In the case of irradiation of a material, the intensity of the release of energy into the electron subsystem of a solid by a fast ion can be 10^3-10^4 times greater than the release of energy into the nuclear subsystem [130]. This difference boosts the role of electronic excitations in the processes of defect generation, causes intense inelastic sputtering of the material [131], including the appearance of a number of specific effects, such as local melting, amorphization and generation of shock waves. The most interesting of the possible results of the passage of a high-energy ion through a solid is the formation of a hidden track. These macrodefects are detected by chemical etching of irradiated material [132, 133]. Experimental studies [134] show that, depending on the value of the energy released by a fast ion into the electronic subsystem of a solid, different morphologies of defect formations can be observed, while a relatively high value of energy leads to the formation of long cylindrical tracks.

Irradiation of the material with high-energy ions is not the only mechanism for creating in the material three-dimensional defects of cylindrical shape. For example, the formation of cylindrical pores was observed when the mother solution exits from the crystal volume to the surface when copper sulphate is heated [135]. In addition, when an electron beam is exposed to the metal surface, a capillary is observed [136].

The study of such defects is of interest due to the fact that the material containing extended nanopores of cylindrical shape can be widely used in the manufacturing of filters, detectors, cooling elements in nano-electronics, and so on.

The purpose of the computer experiments described in this section was to study the effect of shock waves on the structural changes of nanopores of a cylindrical shape.

The simulated crystalline gold originally contained 20,000 atoms. To create a pore in the computational cell, a through hole of a cylindrical shape with a certain radius of the base was created. Then the axis of the cylinder was aligned with one of the lattice sites, and all the atoms entering this cylinder were removed. Since periodic boundary conditions were used in all directions, the result was a cylindrical pore of infinite length. After the removal of the atoms, the structural relaxation of the computational cell was carried out before the system entered a state with a minimum energy, and the resulting structure was used in future studies as the starting one. During the experiments, cylindrical pores of different diameters were considered, with the axes of cylinders oriented in different crystallographic directions. For example, Fig. 3.45 shows a fragment of a computational cell containing a cylindrical pore oriented along the Z axis (this axis corresponds to the crystallographic direction <111>), with a base diameter of 23 Å.

The study showed that the pores of the cylindrical form remain stable at a temperature below $\approx 0.45 \cdot T_m$. When the temperature is raised, structural transformations are observed, consisting in the formation of tetrahedra of packing defects. This phenomenon, as was shown above, is also observed with pores of a spherical shape. In the process of structural relaxation, dislocation loops are generated, the source of which is the inner surface of the pore, and since gold is a metal with a low energy of packing defects, then the splitting of loops occurs and the tetrahedra of packing defects are formed. At the first stage of the transformation, relatively small tetrahedra appeared, having various geometric imperfections, which include truncated tops or steps on the faces. Subsequently, the tetrahedra were enlarged and the total length of their edges decreased, which, in fact, are the nuclei of vertex dislocations. The process of enlargement of packing defect tetrahedra by combining several fragments or by absorbing vacancies occurs as long as this configuration remains energetically feasible. For gold, the dimensions of the lengths of the edges of tetrahedra are limited to 200 Å [137].

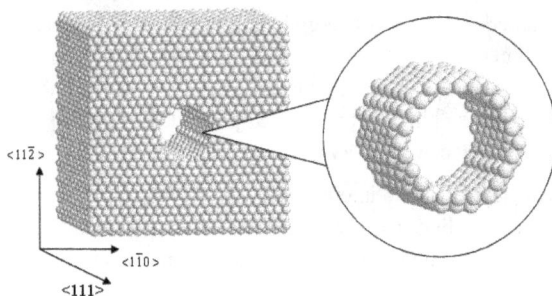

Figure 3.45: Computational cell containing a cylindrical pore. The magnified section shows the surface of the pore.

When the cylindrical pore was oriented along the Z-axis, it was observed a relatively ordered formation of tetrahedra of packing defects along the pore axis in comparison with the orientations along the X and Y axes. The reason for this, in all likelihood, is that the direction of the Z-axis coincides with the normal to the face of the tetrahedra.

Consider the process described above with a specific example. The computational cell containing a cylindrical pore with a base diameter of 14.5 Å was kept for a certain amount of time at a temperature of 900 K. Then, by zeroing the atomic velocities, the structure of the system was fixed and using the visualizer of the distribution of potential energy, the resulting structure was depicted. For clarity, the atoms having a higher energy were coloured in a darker colour. Thus, in Fig. 3.46a, a defect structure is represented, which is fragments of the packing defect tetrahedra formed in the computational cell through 2.5 ps of the computer experiment. As can be seen from Fig. 3.46a, the vertices of the tetrahedra formed are directed in opposite directions (see schematic illustration). The ideality of tetrahedra is disrupted because, as a rule, one of the faces is unfinished. Later, these fragments are combined into a single, larger complex, which is also not a perfect tetrahedron (see Fig. 3.46b). In addition to not completely aligned faces, one of them has a step (see the schematic image).

Analysing Fig. 3.46a, it can be concluded that the dislocation loops generated by the pore are doubles, which is the more preferable configuration. It is known that the energy of a double packing defect is less than the sum of two energies of conventional packing defects, so the probability of double loop formation in a crystal increases [137].

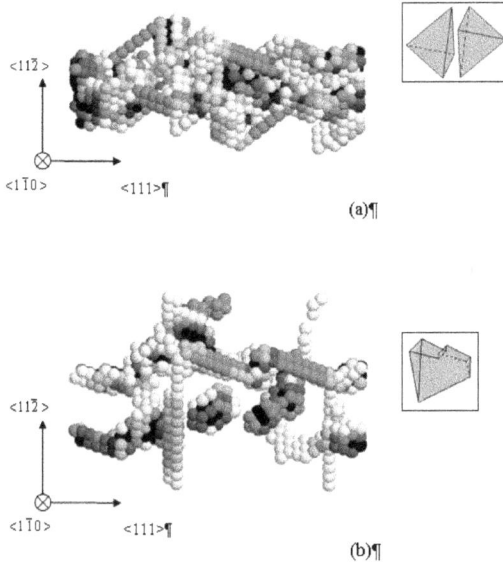

Figure 3.46: *Fragments of the packing defect tetrahedra in a computational cell with a cylindrical pore, observed after 2.5 ps (a) and 25 ps (b) of the computer experiment. The temperature of the computational cell is 900 K. The schematic representations of the tetrahedra of packaging defects are shown on the right-hand side.*

The dislocation loops generated by the pore surface are located in the most densely packed plane of a crystal with an fcc lattice, which is the {111} plane. Therefore, when the axis of the cylindrical pore is oriented along the direction of the computational cell Z, the loops are arranged symmetrically in parallel planes. Their subsequent splitting forms an ordered arrangement of tetrahedra. But the configuration of a twin tetrahedron in the form of a "Christmas tree" is not energetically feasible. To confirm this, we will carry out the following experiment. Consider the ideal tetrahedron of packaging defects. To create it in the {111} plane of the computational cell, we remove the atoms in such a way that an equilateral triangle is formed. In the process of structural relaxation, the created vacancy disk collapses, forming a tetrahedron. Fig. 3.47a shows the energy distribution of atoms in the resulting tetrahedron of packing defects using a visualizer of potential energy. As can be seen from this figure, the atoms with the highest energy are located at the vertices of the tetrahedron, which is quite obvious. We also note that the energy of atoms located on the edges of the tetrahedron is slightly elevated at the edge that was created at the beginning of the experiment by removing atoms. Fig. 3.47b shows the double tetrahedra, from which it follows that the atoms have increased the energy while"crossing" the edge of one of the tetrahedra.

Let us determine the energy of formation of such a configuration of tetrahedra by formula

$$\Delta U = U_{def} - nU_0, \tag{3.5}$$

where U_{def} is the potential energy of a computational cell containing a defective structure, after the structural relaxation, nU_0 is the energy of an ideal computational cell containing the same number of atoms. Calculations show that the energy of formation of this configuration can exceed the energy of formation of two separate tetrahedra by 25%. Of all the possible dual configurations, the most advantageous is that when two tetrahedra of packing defects have one or two common vertices. This was confirmed by calculations in [106], where the tetrahedra of packing defects, formed as a result of the collapse of different vacancy disks, were considered. In addition, if we consider the same number of vacancies, then in the case of the formation of one tetrahedron, the energy gain can be 20%, compared with the formation of two tetrahedra. For comparison, in the model under consideration, the energy of the formation of two separate vacancies exceeds the energy of formation of bivacancy by 7%.

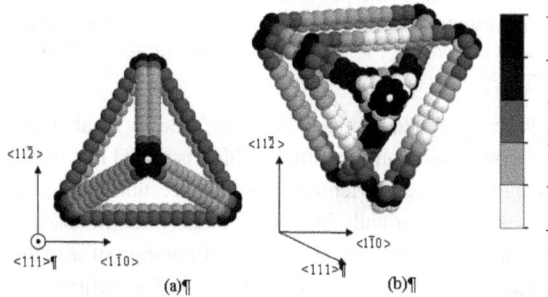

Figure 3.47: *A single (a) and dual (b) tetrahedron of packing defects, represented with the help of the visualizer of the potential energy distribution (EV).*

At the next stage of the study, the computer experiments were considered in which the cooperative atomic displacements were created that simulate the propagation of a wave in a computational cell. The experiments carried out showed that if the wave was assigned to a velocity corresponding to the propagation speed of longitudinal sound waves c_p, then, as a result of the passing of the cylindrical pore with several wavefronts, the tetrahedra of packing defects could be formed at temperatures insufficient for their nucleation as a result of the above-described thermal activation. Thus, Fig. 3.48 shows the defect structure formed as a result of passage through the pore of five waves generated with an interval of 2.5 ps at a temperature of the computational cell of 300 K.

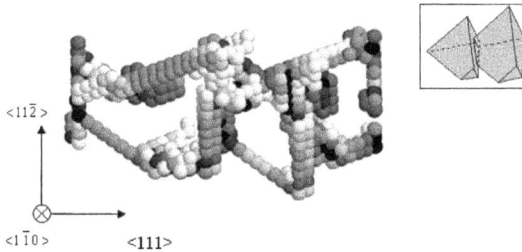

Figure 3.48: *Fragments of packing defects tetrahedra in a computational cell with a cylindrical pore, observed after passing through the cell of five sound waves. The temperature of the cell is 300 K. A schematic representation of the tetrahedra of packaging defects is shown on the right-hand side.*

It can be seen from Fig. 3.48 that as a result of the passing of sound waves through the pore, fragments of double tetrahedra are formed, whose shape ideality is broken due to the incomplete construction of vertices. Thus, the sound waves can initiate the nucleation of dislocation loops, which subsequently lead to the formation of packing defects tetrahedra.

Let us consider the effect of shock waves propagating in the computational cell on the structural transformations of cylindrical pores. Shock waves were created by imparting a velocity of $2 \cdot c_p$ to the group of boundary atoms. The carried out research has shown that with the help of shock waves it is possible to split groups of vacancies from the "parent" pore. The number of splittable vacancies varies depending on the frequency of wave generation. In addition, another way to regulate this number, as has already been shown, is the deformation of the computational cell, by means of which it is possible to split the pore. For this purpose, a computational cell containing a cylindrical pore with a base diameter of 23 Å (see Fig. 3.45) was subjected to uniaxial tension along the Z direction with a strain value $\varepsilon_z = 3\%$. Shock waves were then generated at an interval of 2.5 ps. After passing through three waves, the "division" of the pore into two parts of approximately the same size is observed (see Fig. 3.49a). If in the future we generate waves from the opposite boundary of the computational cell, then as a result of their action on the split-off pore, one of the parts is removed from the second. As a result of subsequent exposure of the computational cell at a temperature of 600 K and cooling, by multiple zeroing of the atomic velocity, the "parent" pore is divided into two separate cylindrical pores (see Fig. 3.49b). As the number of generated waves increases, the distance between the generated pores also increases.

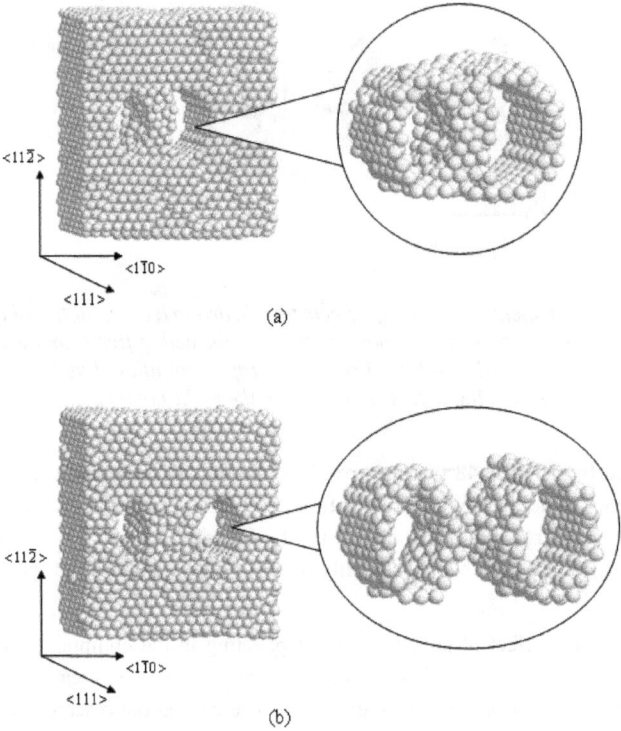

(a)

(b)

Figure 3.49: *The computational cell containing a cylindrical pore, after the passing of three shock waves in the direction* $<1\bar{1}0>$ *(a), and after the passing of the same number of waves in the opposite direction (b). The magnified sections show the surface of the split pore.*

Thus, the conducted study showed that cylindrical nanopores remain stable up to a certain temperature, which for a gold crystal is $\approx$ 600 K. At higher temperatures, the pores collapse, transforming into complexes of packing defect tetrahedra. In the process of structural relaxation, the tetrahedra are enlarged due to the absorption of individual vacancies or neighbouring tetrahedra. Consequently, it can be assumed that in the place of a cylindrical pore of a large extent an ordered system of packing tetrahedra can form.

The critical temperature at which the cylindrical nanopore is transformed can be reduced by the impact of elastic waves on the pore. Under the influence of shock waves, cylindrical nanopores can be split into individual components. In combination with the tensile deformation by means of shock waves, it is possible to divide a cylindrical nanopore into two separate pores of smaller diameter having the same orientation.

The results of the experiments [138-189] testify to the important role of the post-cascade shock waves in the processes of nucleation, growth, and subsequent structural transformations of nanopores. A distinctive feature of the processes described above is the possibility of their occurrence at temperatures insufficient for the onset of diffusion processes. The obtained results can be used, both in radiation materials science and in predicting the behaviour of materials operated under extreme conditions. For example, it is known that the main ways to reduce the radiation swelling of structural materials are to change the structural state of materials by alloying, mechanical and thermal treatments. It is also possible that the results described in this chapter may contribute to the development of a new technique for controlling swelling.

4.   Formation of nanopores in bimetallic particles under the influence of shock waves

4.1   Influence of the post-cascade shock waves on the interphase boundary of bimetals

Various bimetallic compounds are actively used in many fields of science and technology. Wear-resistant and tool bimetals are increasingly used in machine parts and subjected to severe wear [190]. Antifriction bimetals have long been used for the production of sliding bearings. Conducting and contact bimetals are actively used. Bimetals for deep drawing combine, in addition to high strength and sufficient ductility, good thermal conductivity and corrosion resistance [191-192]. In turn, with the transition to nano-dimensional materials, the theoretical and experimental interest in bimetallic nanoparticles has sharply increased. The unique physical and chemical properties of bimetallic particles are associated with their structural, electronic, and optical properties [193].

The structure of bimetallic nanoparticles is determined by the distribution of metals in it. Particles can be organized in the form of an alloy of arbitrary composition or have a core-shell architecture. The latter type is realized only at the nano level and represents particles of the same metal, covered by another. Such materials attract considerable research interest in connection with their potential application in the field of heterogeneous catalysis since they are often more active than their monometallic counterparts. The improvement in the properties of these systems is associated with the complex interaction of electrons of two metals and the effects of lattice parameter changes in bimetallic alloys or on the interphase boundaries of two metals [193-194]. Of special interest are compounds of Ni-Al and Ni-Fe. So bimetallic Ni-Fe nanoparticles are used in the production of hydrogen gas, which isa good catalyst that allows to significantly reduce the level of tar at a smaller surface area [195-196]. Bimetallic Ni-Fe particles act in the catalyst's role in the production of methane [197]. Bimetallic Ni-Al compounds are used to produce multilayer carbon nanotubes from polypropylene [198] and other compounds [199].

Bimetallic compounds, due to their use in various technological processes, can be subjected to various intense external influences, which can lead to energy and structural transformations and which, in turn, affect the properties of such particles. Thus, we see the most relevant study of the effect of the flow of high-energy particles on a solid body, accompanied by the formation of post-cascade shock waves formed as a result of a sharp expansion of a strongly heated cascade region [3] and the passage of such waves of the boundary interface of metals.

The model considered was a two-dimensional crystal [200]. The choice of the dimension of the system is determined by a number of factors. Recently interest in two-dimensional systems has been growing due to the discovery of two-dimensional materials and prospects of their use. Also, two-dimensional models make it possible to visualize more clearly the processes taking place in crystals. In addition, less computer time is required to perform the calculations.

In this section, all experiments on two-dimensional models were carried out for the <111> plane of the fcc lattice. The choice of this plane is due to the fact that this plane is the most closely packed and, consequently, such a model will be the most stable. It is also known that diffusion processes, as a rule, develop in close-packed directions, which correspond to the <111> plane in a conventional fcc crystal. The computational cell represented a rectangle (Fig. 4.1) with sides along the crystallographic directions along the X and Y axes in the crystallographic parameters of the three-dimensional fcc lattice. The number of atoms in the computational cell ranged from 1600 to 4500, depending on the experiment. The atoms interacted through the Morse pair potential:

$$\phi_{PQ}\left(r_{ij}\right)=D_{PQ}\beta_{PQ}\exp\left(-\alpha_{PQ}r_{ij}\right)\left(\beta_{PQ}\exp\left(-\alpha_{PQ}r_{ij}\right)-2\right),\tag{4.1}$$

where D is the binding energy corresponding to the depth of the potential well, α is the parameter that determines the rigidity of the interatomic bonds, r determines the equilibrium interatomic distance. The parameters were calculated by the standard procedure [20] from the following conditions:

$$\frac{1}{2}\sum_{i=1}^{z}\eta_i\varphi_{V=V_0}=E_S,\ \frac{1}{2}\sum_{i=1}^{z}\eta_i\left(\frac{\partial\varphi}{\partial V}\right)_{V=V_0}=0,\ -V_0\cdot\left(\frac{\partial P_S}{\partial V}\right)=K_0.\tag{4.2}$$

Here E_S is the sublimation energy of the crystal atoms at zero temperature, K_0 is the bulk modulus of elasticity, and V_0 is the equilibrium volume.

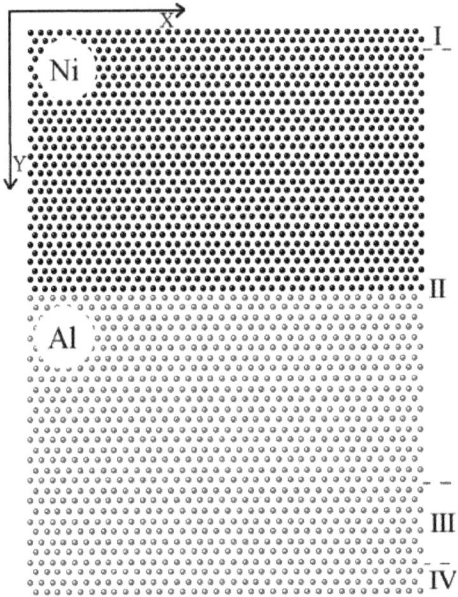

Figure 4.1: *The computational Ni-Al bimetal cell.*

To approximate the real crystals, a boundary condition was imposed on the cell. The periodic boundary conditions along the X-axis were imposed. Along the Y-axis, the boundary conditions were formed as follows. The formation of a post-cascade shock wave (PSW) occurred from the side of Ni (see Fig. 4.1) with free boundary conditions, in Fig. 4.1 this region is marked with the number I. The number II marked the artificially created boundary of the interface of metals, which underwent a relaxation procedure, and during which the boundary atoms occupied an equilibrium position. As a result of the relaxation, an increase in the cell temperature to several tens of Kelvin was observed. The relaxation time of the cell was 100 ps, and the cooling stage was 50 ps. Such a time frame of the experiment is enough for the cell to get rid of the excess free volume that appeared on the interface of metals when creating the initial structure. As a result, an interface with characteristic misfit dislocations was formed (see Fig. 4.2). The buffer zone designed to damp a shock wave that has passed the interface of metals and prevent the formation of a reflected wave from the boundary of the computational cell is marked by the number III. In this region, the velocity of atoms decreases by 5% at each step of integration, i.e. this is the cooling area. The boundary of Al (Fe for Ni-Fe), which represents a region with rigid boundary conditions, is marked by IV.

Figure 4.2: The boundary of Ni-Al bimetal with formed misfit dislocations.

To create a wave, the atoms in the boundary region I (Fig. 4.1) of the computational cell were assigned a velocity commensurate with the speed of sound in the Ni crystal, along the direction.

In a number of works [203-208], it was shown that dynamic crowdions and soliton-type waves can lead, both to the slip of dislocations, and to their climbing, as well as to the directed drift of defects near the misfit dislocations.

When simulating the passage of the PSW through the crystal, the behaviour of the atoms near misfit dislocations was monitored, depending on the velocity of the wave. The velocities were set in the range of 0.7-1.6 of the sound velocity in Ni, for Ni-Al bimetal and 0.7-2 for Ni-Fe bimetal.

Consider the effects that arise near the Ni-Al boundary. As a result of the experiments, it was established that at the initial velocity of the wave 0.7-1.1 of the sound velocity in nickel after its passage through the interface of metals was observed a slip of dislocations along the metal boundary caused by an increase in the temperature of the computational cell. At the same time, we note that the front of the wave experiences distortions during the passage of the boundaries of metals. In fact, the misfit dislocations become sources of secondary waves, which leads to their superposition (interference) and, as a result, to the destruction and release of energy near the metal boundary, mainly in the Al lattice. At velocities of 1.1 - 1.3, the shock waves caused the misfit dislocations to climb into the Al lattice, which cannot be explained by an increase in the temperature of the computational cell to 300 K.

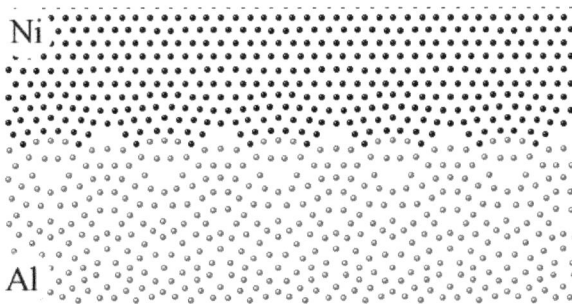

Figure 4.3: The boundary of Ni-Al bimetal at the time of the passing of the shock wave, with characteristic regions of compression and expansion near the metal boundary.

The phenomenon of interference of secondary waves after passing through the bimetal boundary was most clearly observed at initial velocities of the shock wave 1.3 - 1.6 of the sound velocity in Ni. This is due to the decrease in the wavelength, as well as the increase in the energy transferred by it. As a result of the passage of the misfit dislocations by the shock wave, regions with an increased and decreased concentration of atoms (diffraction maxima and minima, Fig. 4.3) were formed in the aluminium lattice.

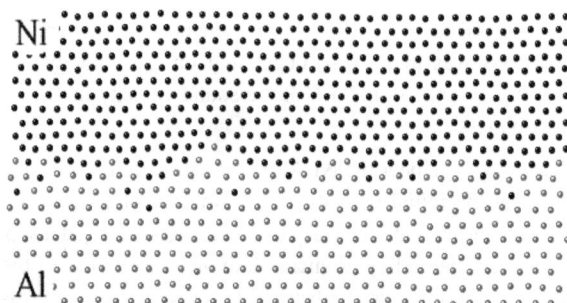

Figure 4.4: *The structure of the Ni-Al bimetal boundary after the passing of the shock wave at a speed exceeding 1.4 times the sound velocity in Ni.*

The energy release in the border zone of the bimetal led to the heating of the Al lattice to temperatures close to the melting point, as a result of which the diffusion mechanisms became more active and the characteristic boundary of the metals was formed (see Fig. 4.4). At a shock wave velocity of 1.5-1.6 and more than the sound velocity in nickel, pore nuclei in the near-boundary region on the Al side were formed (Fig. 4.5). The nuclei often collapsed with the formation of a dislocation loop, or they could develop into a stable existing pore.

Similar experiments were carried out with Ni-Fe bimetal. To analyse the differences in the results of the experiments performed for Ni-Al and Ni-Fe, some of their parameters are given in Table 4.1.

Due to the proximity of the given physical parameters of Ni and Fe, the effects that were clearly manifested for the Ni and Al boundaries were less pronounced here. Because of the smaller differences in the lattice parameter, the misfit dislocation density of Ni-Fe was less than Ni-Al. This reflected on the number of compression/tension regions near the boundary and increased the distance between them. It is established that for the climb of misfit dislocations in Ni-Fe, a shock wave with a velocity of at least 1.5 sound velocities in nickel is necessary, and for the formation of pore nuclei, 1.9 sound velocities in nickel and more are required. Such dependencies are also due to the melting points of the bimetal components. If Al has a melting point 1.85 times lower than that of Ni, then iron has a slightly higher melting point than Ni. The energy of the shock wave, when scattered near the boundary of the bimetal, leads to local heating, and in the case of Al, to the actual melt and rearrangement of the dislocation structure, more energy is required for iron to activate these processes.

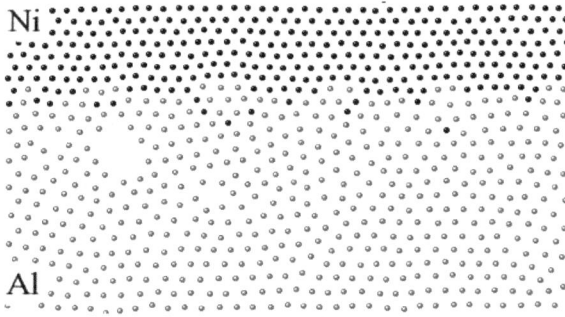

Figure 4.5: *Formation of the pore near the Ni-Al bimetal boundary after the passing of the shock wave at a speed exceeding 1.6 times the sound velocity in Ni.*

Table 4.1 Physical characteristics of bimetals.

Type of bimetal	Ratio of nominal sizes of atoms	Ratio of nominal masses of the components	Ratio of the elastic moduli of the components	Ratio of melting points of the components
Ni – Al	0,87	2,17	2,87	1,85
Ni – Fe	0,9	1,05	0,99	0,95

The studies carried out by the molecular dynamics of the influence of shock waves on the structure of the model boundary of Ni-Al and Ni-Fe bimetals have shown that, depending on the initial velocity of the wave motion, it can cause misfit dislocation slip, their climbing, and also lead to the formation of pores near the bimetal boundary . Such structural changes can affect the physical and chemical properties of bimetallic particles.

4.2   The passage of shock waves through the interface of bimetallic particles

The processes considered in the previous section will undoubtedly be influenced by the finite dimensions of the bimetallic particles and the free surfaces of the particles. Therefore, in what follows we will consider a three-dimensional model of bimetallic particles of different sizes and mutual orientations. The study is carried out with the example of Ni-Al bimetal since the effects obtained in it were most pronounced.

The simulation was carried out using the LAMMPS MD-modelling package, the obvious advantage of which is a wide range of supported potentials, comparative ease of use and the open source code. This package was developed for application to calculations on parallel computers.

As a potential function of interatomic interaction, the potential included in the standard set of LAMMPS calculated in the framework of the immersed atom method (EAM) was used. The temperature of the computational cell was set by assigning random velocities to atoms in accordance with the Maxwell-Boltzmann distribution for this temperature. The step of numerical integration of the equations of motion was equal to 1 fs.

In the present work, the study was carried out on a computational cell simulating the Ni-Al bimetal (Fig. 4.6), which had the form of a rectangular parallelepiped, the number of atoms and the cell size varied to determine the influence of the size factor on the behaviour of the bimetal during the passage of the shock wave. The minimum linear dimensions of the computational cell were 6.6 nm along each edge of the parallelepiped or 4.5×10^3 atoms in the cell, the models whose linear dimensions reached 65 nm along one of the edges of the parallelepiped containing about 2.3×10^6 particles were considered maximally. The thickness of the Al layer varied from 4.2 to 15.45 nm.

To obtain model bimetals, two initial single-component crystals of different metals in the form of rectangular parallelepipeds were placed at a distance of about 2.5 Å from each other. After that, the structure was relaxed. Along all axes, free boundary conditions were set.

Three orientations of bimetallic particles in space were considered:

I. The X-axis was directed along the crystallographic direction $< 10\bar{1} >$, the Y-axis was along <010>, and the Z-axis was along <101>;

II. The X-axis was directed along the crystallographic direction <100>, the Y-axis was along <010>, and the Z-axis was along <001>;

III. The X-axis was directed along the crystallographic direction $< 1\bar{1}0 >$, the Y-axis was along $< 11\bar{2} >$, and the Z-axis was along <111>.

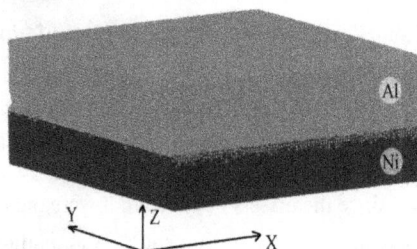

Figure 4.6: *The 3-D view of a bimetallic Ni-Al block containing 834,000 atoms.*

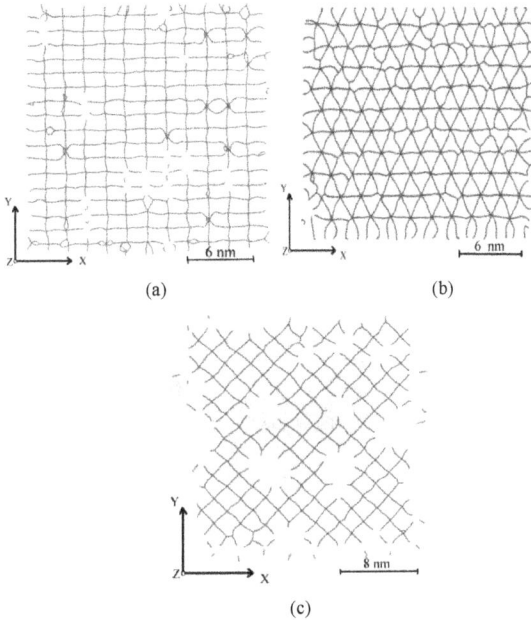

(a)   (b)

(c)

Figure 4.7: *The misfit dislocation grid at the interface of bipartite bimetallic Ni-Al particles, (a) the X-axis is directed along the crystallographic direction <010>, the Y-axis is along $< 10\bar{1} >$ and the Z-axis <101>; (b) the X-axis is directed along the crystallographic direction <100>, the Y-axis is along <010>, the Z-axis <001>, (c) the X-axis is directed along the crystallographic direction <111>, the Y-axis is along $< 11\bar{2} >$, and $Z < 1\bar{1}0 >$.*

At the metal interface, a mesh of misfit dislocations formed due to the difference in the lattice parameters of the components of bimetallic particles and the orientation of its components in the space (Fig. 4.7). Such a configuration allowed us to consider the interface with different density of misfit dislocations and the direction of close-packed atomic rows.

To create the wave, a group of atoms in the border region of the computational cell (one outer atomic layer) was assigned a velocity exceeding the speed of sound waves from the material along the crystallographic direction of the corresponding Z axis. The velocity of atoms was assigned in the interval from 1 c to 1.5 c. Thus, a shock wave was formed that passed through the Ni-Al bimetal interface.

(a)   (b)

(c)

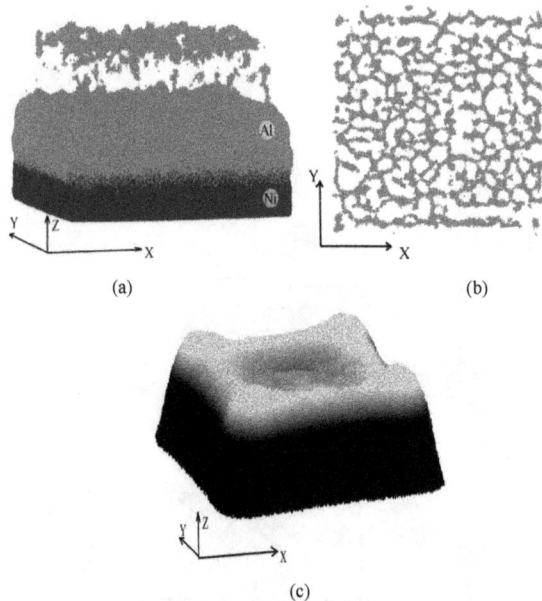

Figure 4.8: *The bimetallic Ni-Al particle (834,000 atoms) oriented for the case I; (a) the 3D view of the Ni-Al nanoclusters after the passage of the shock wave; (b) the visualization of the formed Al clusters; (c) the surface relief following the separation of Al nanoclusters after 5 ps of relaxation.*

Particular attention was paid to the shock wave propagating along the close-packed direction, since in this case, the mechanism of focusing of the energy takes place, and the spherical wave is transformed into fragments of plane waves propagating precisely along the close-packed directions.

As shown in the previous section, the passage of shock waves is accompanied by their diffraction on the misfit dislocation nuclei near the metal interface. With sufficient wave energy, the nuclei of the pores in Al were formed. The misfit dislocation grid for the three-dimensional case has a complex structure, so the distribution of the interference maxima (minima) also has a more complex form.

We consider two directions of propagation of shock waves. In the first, the wave was initiated from the side of Ni. Its velocity varied from 48 Å/ps to 72 Å/ps, in accordance with the tabulated value of the sound velocity in Ni crystallite [209].

In the case of orientation I, the wave propagation direction corresponded to the close-packed direction in the crystal. With this configuration, the wave passes through the interface of metals with minimal distortion of the front and energy losses (Fig. 4.8a). The evaluation showed that about 45% of the energy of the initial wave comes up to the surface of Al, while the outer Al atoms break off to form nanoclusters, whose initial dimensions correspond to the linear dimensions of the mesh dislocation grid cells (Fig. 4.8b). After the split of Al atoms, the characteristic surface of the bimetallic particle was formed (Fig. 4.8c). The shape of the surface is due to the presence of free boundary conditions.

At shock wave velocities of more than 60 Å/ps, pore nuclei were formed near the interface of the bimetallic particle on the Al side. Further relaxation of the structure led to the collapse of the formed pores. The evolution of the porous structure is fairly rapid. The main trend is to combine the pores from the edges to the centre and then collapse.

In more detail, the evolution of the pores will be considered for the II and III cases of the orientation of the bimetal components. For the entire velocity interval, the formation of pore nuclei near the metal boundary took place. As experiments showed, the initial velocity of the shock wave affected the size of the pore nuclei that formed on the Al side (Fig. 4.9).

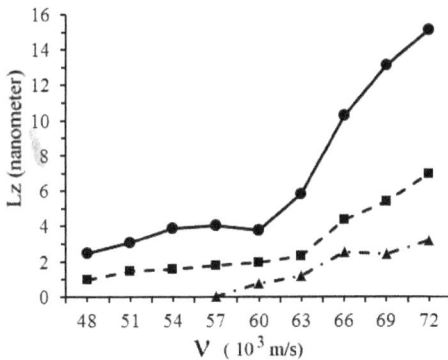

Figure 4.9: *Dependence of the linear size of the pore along the Z-axis on the initial velocity of the shock wave, the round marker corresponds to the direction of propagation of the wave <111>, the square one <001> and triangular <101>. The linear dimensions of the bimetal along the X and Y axes are 41.8 nm and along the Z-axis 14.1 nm.*

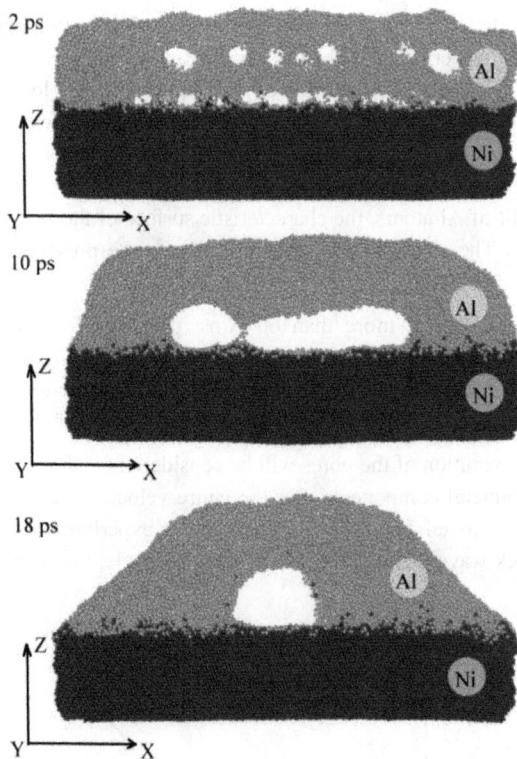

Figure 4.10: *Evolution of pores in a bimetallic particle of orientation III after the passing of the shock wave with a velocity of 60 Å/ps. The thickness of the visualized plane is 4 nm. The linear dimensions of the bimetal along the X and Y axes are 41.8 nm and along the Z axis 14.1 nm.*

The pore nuclei for the case III were significantly larger than for other orientations of the bimetal components at the same initial wave velocity since there was a denser presence of misfit dislocations at the metal interface. This was most clearly manifested for velocities of more than 60 Å/ps. Also, the dissipation of energy of the shock wave after passing through the boundary was facilitated by a non-closely packed direction in the crystal.

In this case, the pore distribution and the rate of their collapse in bimetal depended on the linear dimensions of the particle. Note that for particles of the dimension less than 8 nm the pores were not formed. This is due to the influence of the free surface of the bimetal and to a smaller number of dislocations at the interphase boundary. For bimetals measuring 8 to 15 nm, two or three pores are formed in the regions of the first and second interference minimum of the shock wave on misfit dislocations. For larger cell sizes, the number of pore nuclei increased, an example of the evolution of this structure is shown in Fig. 4.10.

The size of the bimetallic particles influenced the pore collapse time. Fig. 4.11 shows such a dependence for the orientation of the components of bimetal (case III) from the linear cell dimensions along the X and Y axes. The thickness of the Al layer was a constant value of 5.84 nm for all the experiments.

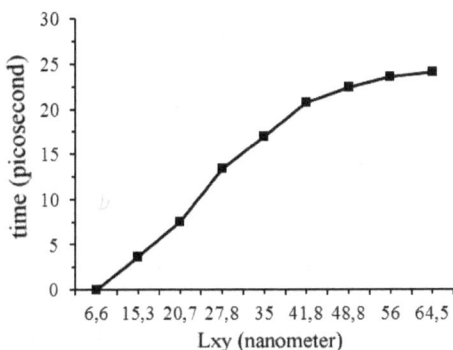

Figure 4.11: Dependence of pore collapse time on the linear dimensions of a bimetallic particle along the X and Y axes for the case III.

The results obtained indicate that the effect of the free surface decreases with increasing of linear dimensions of the cell. It is important for these processes even for the particles greater than 60 nm in dimension.

Next, the effect of the thickness of the Al layer on the possibility of pore formation was considered, the results are shown in Fig. 4.12. The linear dimensions of the cell along the X and Y axes were kept at 41.8 nm. The results obtained indicate that at a thickness of more than 15 nm, the pore formation does not occur. Despite the Al melt near the metal interface, the energy of the wave is not sufficient to create a free volume due to the displacement of Al atoms along the Z-axis.

Thus, the observed effects are largely determined by the dimensions of the bimetallic particles.

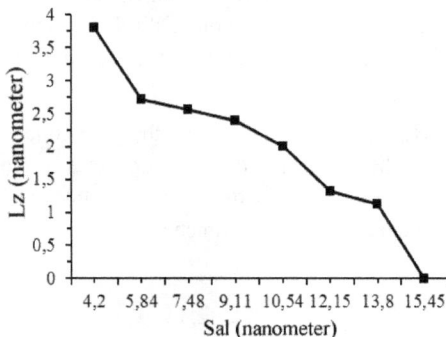

Figure 4.12: *Dependence of the linear pore sizes along the Z-axis on the thickness of the Al layer, the linear dimensions along the X and Y axes are 41.8 nm.*

In the case of the initiation of a shock wave on the Al side, the wave velocity was set in the range from 50.8 Å/ps to 76.2 Å/ps, in accordance with the sound velocity in Al [209]. The obtained results show that, despite the change in the direction of propagation of the shock wave, the pore nuclei were formed on the Al side. This is due to the partial reflection of the shock wave from the metal boundary and the lower binding energy of the Al atoms to each other. The orientation of the components of the bimetal influenced the size of the nuclei of the pores in a similar manner, as in the initiation of a wave from the side of Ni. In the case of the propagation of a shock wave along a close-packed direction, the pore size was minimal. Thus, regardless of the direction of the shock wave, it is possible to form pore nuclei only on the Al side near the metal interface.

Computer simulation of the passage of a shock wave through the interface of a bimetallic Ni-Al particle, carried out by the molecular dynamics method, showed a significant effect on the structure of the boundary and the surface of the crystal. It has been established that one of the determining factors for the formation of pore nuclei near the Ni-Al interface is the mutual orientation of the metals, their size and the direction of the passage of the shock wave. In the propagation of a wave initiated on the Ni side along a close-packed direction, apart from the formation of pores, it is possible to split a part of the atoms from the Al surface and then form clusters from them. This is due to the fact that the front of the shock wave undergoes minimal distortions during the passage of the metal interface, and a significant part of the wave energy reaches the surface of the Al crystal. In the case of crystal orientation, when the wave does not propagate along the close-packed direction, the bulk of the energy is scattered near the interface of the metals and leads to the formation of pore nuclei. The formation of pores depends more on the particle size of the bimetal. In the case of the initiation of a shock wave on the Al side, similar effects were observed, however, the pores were formed on the Al side, as in the previous case. This arrangement of pores is due to a lower binding energy between Al atoms than between Ni-Ni and Ni-Al.

Conclusions

To date, the problem of designing radiation-resistant materials is becoming important in connection with the widespread use of nuclear power sources and the rapid development of the rocket and space industry. The irradiation to which the applied structural materials are subjected inevitably leads to a modification of their structure, and as a result, a change in the strength, electro-physical and other properties. The study of structural transformations in metals and alloys, when irradiated with fluxes of charged particles capable of initiating various processes of atomic tuning, is of great interest, since the mechanism of energy transfer (elastic or inelastic interaction, ionization) can be purposefully changed by choosing the type and energy of the irradiating particles, which opens up wide prospects for using radiation exposure as a tool for technological processing.

In the framework of the molecular dynamics study, the high-speed cooperative displacements of groups of atoms in metals with an fcc lattice that arise upon the radiation effect on the crystal structure and their influence on the structural transformations of nanopores were considered. As one of the possible mechanisms for the homogeneous nucleation of nanopores, the following mechanism is proposed. As a result of radiation exposure, the shock waves are formed, which, propagating in the crystal, can cause a drift of vacancies. Spontaneously formed vacancy clusters after the passage of the shock wave are transformed into a complex that is one or more tetrahedra of packing defects. The study showed that these data are the most stable, and subsequent waves do not cause their transformation. Small vacancy clusters do not represent an insuperable obstacle for such waves, therefore single vacancies can be attracted to such clusters as a result of the passing of several shock waves, even at temperatures insufficient for the onset of diffusion processes. Homogeneous formation of the pores is simplified due to the tensile stresses arising in the crystal after the passage of the shock wave. In addition, the study showed that shock waves affect structural changes in nanopores, and can cause their displacement, as well as splitting or dissolution. In this case, dissolution can be carried out even at temperatures insufficient to initiate the thermal activation of the diffusion dissolution mechanism. In addition, with a certain arrangement of the pores, the shock waves can initiate their coarsening, even at relatively low temperatures of the crystal, which are insufficient for the beginning of so-called heat-induced coalescence. The processes of structural transformations are affected by the number of vacancies, as well as the temperature at which these processes occur. A study of the effect of the deformation of the computational cell showed that the role of the surface energy of the pore increases with compression so that the waves activate processes that in some cases can lead to the fusion of individual pores. The tension leads to the stabilization of the pores.

The study shows that high-speed cooperative atomic displacements are a powerful tool that allows the purposeful modification of defective structures of crystalline bodies. The results obtained may possibly be used in radiation material science, nano-engineering, or in the ultrasonic treatment of materials.

References

[1] Workshop on advanced computational materials science: application to fusion and generation IV fission reactors, March 13 - April 2, 2004. Accessed August 1, 2018. http:// www.csm.ornl.gov/meetings/SCNEworkshop/DC-index.html.

[2] Voevodin, V.N. "Structural materials of nuclear power engineering – the challenge of the 21st century." *Questions of Atomic Science and Technology*, no. 2 (2007). 10-22, [in Russian].

[3] Ovchinnikov, V.V. "Radiation-dynamic effects. Possibilities for the formation of unique structural states and properties of condensed media." *Achievements in the Physical Sciences* 178, no. 9 (2008). 91-1001, [in Russian].

[4] Shtremel, M.A. *Strength of alloys. Part I. Defects of the lattice*. Moscow: National University of Science and Technology MISIS, 1999, [in Russian].

[5] Devyatko, Yu.N., S.V. Rogozhkin, and V.M. Chernov. "Correlation effects in cascades of atomic-atomic collisions." *Questions of Atomic Science and Technology* 65, no. 2 (2005). 132-137, [in Russian].

[6] Biersack, J.P., and L.G. Haggmark. "A Monte Carlo computer program for the transport of energetic ions in amorphous targets." *Nuclear Instruments and Methods* 174. no. 1-2 (1980). 257-269. https://doi.org/10.1016/0029-554X(80)90440-1

[7] Makhinko, F.F. (2014). Recovery of the plasticity of aluminium alloys using dynamic long-range effects during ion bombardment (Doctoral dissertation).

Retrieved from
http://iep7.iep.uran.ru/uploads/diss/mahinko2014/%D0%94%D0%B8%D1%81
%D0%B5%D1%80_%D0%9C%D0%B0%D1%85%D0%B8%D0%BD%D1%
8C%D0%BA%D0%BE.pdf, [in Russian].

[8] Kalin, B.A. *Physical Materials Science*. Moscow: Moscow Engineering Physics Institute, 2008, [in Russian].

[9] Garber, R.I., and A.I. Fedorenko. "Focusing of atomic collisions in crystals." *Achievements in the Physical Sciences* 83, no. 3 (1964). 385-432, [in Russian]. https://doi.org/10.3367/UFNr.0083.196407a.0385

[10]   Chudinov, V.G., R.M.J. Cotterill, and V.V. Andreev. "Kinetics of the diffuse processes within a cascade region in the sub-threshold stages of fcc and hcp metals." *Physica Status Solidi (a)* 122. no. 1 (1990). 111-120. https://doi.org/10.1002/pssa.2211220110

[11]   Zhukov, V.P., and A.V. Demidov. "Calculation of the displacement peaks in the continuum approximation." *Atomic Energy* 59. (1985). 568-573. https://doi.org/10.1007/BF01122180

[12]   Zhukov, V.P., and A.A. Boldin. "Elastic-wave generation in the evolution of displacement peaks." *Atomic Energy* 68. (1987). 884-889. https://doi.org/10.1007/BF01126098

[13]   Bleikher, G.A., V.P. Krivobokov, and O.V. Pashchenko. *Heat and mass transfer in a solid body under the influence of powerful beams of charged particles*. Novosibirsk: Science, 1999, [in Russian].

[14]   Gushchina, N.V. (2008) An investigation of the regularities of phase and structural transformations in aluminium-based alloys upon irradiation with ions of medium energies (Doctoral dissertation). Retrieved from http:// http://www.iep.uran.ru/russian/diss/af_Guschina.pdf, [in Russian].

[15]   Mozharovsky, S.M. (2011) Development of ion-beam methods for modifying the properties of aluminium alloys on the basis of the use of radiation-dynamic effects (Doctoral dissertation). Retrieved from http:// http://earchive.tpu.ru/bitstream/11683/6715/1/thesis_tpu-2011-70.pdf, [in Russian].

[16]   Cawthorne, C., and E.J. Fulton. "Voids in irradiated stainless steel." *Nature* 216. (1967). 575-576. https://doi.org/10.1038/216575a0

[17]   Ibragimov, Sh.Sh., V.V. Kirsanov, and Yu.S. Pyatiletov. *Radiation damage to metals and alloys*. Moscow: Energoatomizdat, 1985, [in Russian].

[18]   Shalaev, A.M. *Radiation-stimulated processes in metals*. Moscow: Energoatomizdat, 1988, [in Russian].

[19]   Kelly, B. *Radiation damage to solids*. Moscow: Atomizdat, 1970, [in Russian].

[20]   Slezov, V.V., and V.V. Sagalovich. "Diffusion decay of solid solutions." *Achievements in the Physical Sciences* 151. no. 1 (1987). 67-104, [in Russian]. https://doi.org/10.3367/UFNr.0151.198701c.0067

[21]   Ostapchuk, P.N. "Generalized coalescence formalism in the theory of vacancy swelling of metals." *Questions of Atomic Science and Technology* 78. no. 2 (2012). 9-15, [in Russian].

[22]   Geguzin, Ya.E., and N.N. Ovcharenko. "Surface energy and processes on the surface of solids." *Achievements in the Physical Sciences* 76. no. 2 (1962). 283-328, [in Russian].

[23]   Dubinko, V.I., P.N. Ostapchuk, and V.V. Slezov. "The theory of diffusion evolution of an ensemble of pores in crystals under irradiation." *Physics of Metals and Metallurgy* 65. no. 1 (1988). 32-43, [in Russian].

[24]   Kirsanov, V.V. "Radiation defects and related effects." *Soros Educational Journal* 17. no. 10 (2001). 88-94, [in Russian].

[25]   Zelensky, V.F., I.M. Nekhlyudov, and T.P. Chernyaeva. *Radiation defects and swelling of metals*. Kyiv: Naukova Dumka, 1988, [in Russian].

[26]   Parshin, A.M. *Structure and radiation swelling of steels and alloys*. Moscow: Energoatomizdat, 1983, [in Russian].

[27]   Parshin, A.M. Structure, strength and radiation damage of corrosion-resistant steels and alloys. Chelyabinsk: Metallurgy, 1988, [in Russian].

[28]   Orlov, V.L., A.G. Malyshkina, and A.V. Orlov. "The pore formation in metals under radiation exposure." *Bulletin of Tomsk Polytechnic University* 305. no. 3 (2002). 314-318, [in Russian].

[29]   Orlov, V.L., A.V. Orlov, and A.G. Malyshkina. "The formation of nanometer-sized ordered structures of radiation pores in metals." *Izvestiya VUZov. Physics*. no. 2 (2003). 31-35, [in Russian].

[30]   Trushin, Yu.V. "The influence of precipitates of the secondary phase on the radiation swelling of decomposing solid solutions. Part I.: General theory." *Journal of Technical Physics* 62. no. 4 (1992). 1-12, [in Russian].

[31]   Al-Samavi, A. Kh. (2004) Radiation swelling of metals (Doctoral dissertation). Retrieved from http://static.freereferats.ru/_avtoreferats/01002630792.pdf, [in Russian].

[32]   Orlov, V.L., A.V. Orlov, A.Kh. Al-Samavi, and A.A. Grebenkov. "The influence of alloying on the swelling of nickel alloys." *Fundamental Problems of Modern Materials Science.* no. 2 (2004). 70-74, [in Russian].

[33]   Maziasz, P.J. "The precipitation response of 20%-cold-worked type 316 stainless steel to simulated fusion irradiation." *Journal of Nuclear Materials* 85-86. (1979). 713-717. https://doi.org/10.1016/0022-3115(79)90344-1

[34]   Cole, J.L., and T.R. Allen. "Microstructural changes induced by post-irradiation annealing of neutronirradiate austenitic stainless steel." *Journal of Nuclear Materials* 283-287. (2000). 329-333. https://doi.org/10.1016/S0022-3115(00)00072-6

[35]   Tsai, K.V., O.P. Maksimkin, and L.G. Turubarova. "Evolution of secondary-phase precipitates during annealing of the 12Kh18N9T steel irradiated with neutrons to a dose of 5 dpa." *The Physics of Metals and Metallography* 103. no. 3 (2007). 284-291. https://doi.org/10.1134/S0031918X0703009X

[36]   Gubaydulin, A.A., D.N. Dudko, and S.F. Urmancheev. "Simulation of the interaction of an air shock wave with a porous screen." *Physics of Combustion and Explosion* 36. no. 4 (2000). 87-96, [in Russian]. https://doi.org/10.1007/BF02699481

[37]   Charakhchyan, A.A., K.V. Khishchenko, V.V. Milyavsky, V.E. Fortov, A.A. Frolova, I.V. Lomonosov, and L.V. Shurshalov. "A numerical study of convergent shock waves in porous media." *Journal of Technical Physics* 75. no. 8 (2005). 15-25, [in Russian]. https://doi.org/10.1134/1.2014526

[38]   Cheremskaya, P.G., V.V. Slezov, and V.I. Betekhtin. *Pores in a solid body.* Moscow: Energoatomizdat, 1990, [in Russian].

[39]   Palatnik, LS, P.G. Cheremskoy, and M.Ya. Fuks. *Pores in the films.* Moscow: Energoatomizdat, 1982, [in Russian].

[40]   Kirsanov, V.V., and A.N. Orlov. "A computer simulation of atomic configurations of defects in metals." *Achievements in the Physical Sciences* 142. no. 2 (1984). 219-264, [in Russian]. https://doi.org/10.1070/PU1984v027n02ABEH004029

[41]   Agranovich, V.M., and V.V. Kirsanov. "Problems of the modelling of radiation damage in crystals." *Achievements in the Physical Sciences* 118. no. 1

(1976). 3-51, [in Russian].
https://doi.org/10.1070/PU1976v019n01ABEH005118

[42]   Kulesh, M.A. Machine modelling of deformation properties of metals. Perm: PGTU, 1997, [in Russian].

[43]   Bieler, D.R. "The role of machine experiments in the study of materials." In *Machine Simulation in the Study of Materials*, 31-250. Moscow: Mir, 1974, [in Russian].

[44]   Krivtsov, A.M., and N.V. Krivtsova. "The particle method and its use in the mechanics of a deformable solid." *Far-East Mathematical Journal, Far East Division, Russian Academy of Sciences* 13. no. 2 (2002). 254-276, [in Russian].

[45]   Psakhie, S.G., S.I. Negruskul, and K.P. Zolnikov. "Discrete computer models of condensed media with internal structure." *Physical Mesomechanics and Computer Design of Materials* 2. (1995). 77-99, [in Russian].

[46]   Plishkin, Yu.M. "Methods of computer modelling in the theory of crystal defects." In *Defects in crystals and their simulation on a computer*, 77-99. Leningrad: Nauka, 1980, [in Russian].

[47]   Alder, B.J., and T.E. Wainwright. "Phase transition for a hard sphere system." *Journal of Chemical Physics* 27. (1957). 1208-1209. https://doi.org/10.1063/1.1743957

[48]   Nazarov, A.A., and R.R. Mulyukov. *Atomistic modelling of materials, nanostructures and processes of nanotechnology*. Ufa: Publishing Department, Bashkir State University, 2010, [in Russian].

[49]   Poletaev, G.M., and M.D. Starostenkov. "Dynamic collective displacements of atoms in metals and their role in the vacancy diffusion mechanism." *Physics of the Solid State* 51. no. 4 (2009). 686-691, [in Russian]. https://doi.org/10.1134/S106378340904012X

[50]   Dudnik, E.A., and M.D. Starostenkov. *Computer modelling of structural-energy transformations in a two-dimensional crystal*. Barnaul: Publishing House of Altai University, 2005, [in Russian].

[51]   Varaksin, A.N., and V.S. Koziajchev. "Diffusion of hydrogen in palladium: modelling by a molecular dynamics method." *Physics of Metals and Metallurgy*. no. 2 (1991). 45-51, [in Russian].

[52]   Varaksin, A.N., B.G. Polosukhin, and V.S. Koziajchev. "The study of migration of hydrogen in niobium using a computer simulation method." *Physics of Metals and Metallurgy*. no. 7 (1990). 13-19, [in Russian].

[53] Gillar, M.J., J.H. Harding, and R.J. Tarento. "Calculation of defect migration rates by molecular dynamics simulation." *Journal of Physics and Chemistry Solids* 20. no. 16 (1987). 2331-2346. https://doi.org/10.1088/0022-3719/20/16/009

[54] Rakitin, R.Yu., G.M. Poletaev, M.C. Aksenov, and M.D. Starostenkov. "The diffusion mechanisms arising along the grain boundaries in two-dimensional metals." *Letters to the Journal of Technical Physics* 31. no. 15 (2005). 44-48, [in Russian]. https://doi.org/10.1134/1.2035354

[55] Upmanyu M., R.W. Smith, and D.J. Srolovitz. "Atomistic simulation of curvature driven grain boundary migration." *Interface Science*. no. 6 (1998). 41-58. https://doi.org/10.1023/A:1008608418845

[56] Galashev, A.E., and I.G. Mukhina. "Molecular-dynamic modelling of thermal destruction of a bcc crystal." *Physics of Metals and Metallurgy*. no. 12 (1992). 3-10, [in Russian].

[57] Kirsanov, V.V., and Yu.S. Pyatiletov. "Investigation by the method of molecular dynamics of the propagation of cracks in metals and their interaction with point defects." *Physics of Metals and Metallurgy*. no. 8 (1991). 51-58, [in Russian].

[58] Kirsanov, V.V., and Yu.S. Pyatiletov. "Influence of temperature on the propagation of cracks and their interaction with impurity helium atoms in metals with a bcc lattice (investigation by a molecular dynamics method)." *Physics of Metals and Metal Science*. no. 6 (1992). 36-42, [in Russian].

[59] Gumbsch, P., S.J. Zhou, and B.L. Holian. "Molecular dynamics investigation of dynamic crack stability." *Physical Review B* 55. no. 6 (1997). 3445-3455. https://doi.org/10.1103/PhysRevB.55.3445

[60] Rahman, A. "Molecular dynamics studies of structural transformation in solids." *Material Science Forum* 1. (1984). 211-222. https://doi.org/10.4028/www.scientific.net/MSF.1.211

[61] Kashirin, V.B., and E.V. Kozlov. "The influence of interaction potential on the structure and properties of modelled amorphous structures." *Physics of Metals and Metallography* 76. no. 1 (1993). 19-27, [in Russian].

[62] Likhachev, V.A., and V.E. Shudegov. *Principles of the organization of amorphous structures*. St. Petersburg: Publishing house of St. Petersburg University, 1999 [in Russian].

[63]   Lee, B.-J., C.S. Lee, and J.C. Lee. "Stress induced crystallization of amorphous materials and mechanical properties of nanocrystalline materials: a molecular dynamics simulation study." *Acta Materialia* 51. (2003). 6233-240. https://doi.org/10.1016/S1359-6454(03)00446-4

[64]   Golovenko, Zh.V., S.L. Gafner, and Ya.Ya.Gafner. "Investigation of structural states of gold nanoclusters by a molecular dynamics method." *Izvestiya VUZov. Physics* 51. no. 11/3 (2008). 186-190, [in Russian].

[65]   Lummen, N., and T. Kraska. "Investigation of the formation of iron nanoparticles from the gas phase by molecular dynamics simulations." *Nanotechnology* 15. (2004). 525-533. https://doi.org/10.1088/0957-4484/15/5/021

[66]   Polukhin, V.A., V.F. Ukhov, and M.M. Dzugutov. *Computer simulation of the dynamics and structure of liquid crystals.* Moscow: Science, 1981, [in Russian].

[67]   Mazhukin, V.I., and A.V. Shapranov. "Mathematical modelling of heating and melting of metals. Part I. Model and computational algorithm." *Preprint of M.V. Keldysh Institute of Applied Mathematics.* no. 31 (2012), [in Russian].

[68]   Regel, AR., and V.M. Glazov. *Regularities in the formation of the structure of electronic melts.* Moscow: Science, 1982, [in Russian].

[69]   Scholtz, N.N. "Processes in the course of radiation damage in crystals." In *Machine Simulation in the Study of Materials,* 286-350. Moscow: Mir, 1974, [in Russian].

[70]   Verlet, L. "Computer "experiments" on classical fluids. I. Thermodynamical properties of Lennard-Jones molecules." *Physical Review* 159. no. 1 (1967). 98-103. https://doi.org/10.1103/PhysRev.159.98

[71]   Schweizer, S., C. Elsasser, K. Hummler, and M. Fahule. "Ab initio calculation of stacking fault energies in noble metals." *Physical Review B* 46. no. 21 (1992). 14270-14273. https://doi.org/10.1103/PhysRevB.46.14270

[72]   Xu, J., W. Lin, and A.J. Freeman. "Twin-boundary and stacking-fault energies in Al and Pd." *Physical Review B* 43. no. 3 (1991). 2018-2024. https://doi.org/10.1103/PhysRevB.43.2018

[73]   Morris, J.R., J.J. Je, K.M. Ho, and C.T. Chan. "A first-principles study of compression twins in h.c.p. zirconium." *Philosophical Magazine Letters* 69. no. 4 (1994). 189-195. https://doi.org/10.1080/09500839408241591

[74]   Resongaard, N.M., and H.L. Skriver. "Ab initio study of antiphase boundaries and stacking faults in L12 and DO22 compounds." *Physical Review B* 50. no. 7 (1994). 4848-4858. https://doi.org/10.1103/PhysRevB.50.4848

[75]   Tang, S., A.J. Freeman, and G.B. Olson. "Phosphorus-induced relaxation in an iron grain boundary: A cluster-model study." *Physical Review B* 47. no. 5 (1993). 2441-2445. https://doi.org/10.1103/PhysRevB.47.2441

[76]   Sob, M., I. Turek, and V. Vitek. "Application of surface ab initio methods to studies of electronic structure and atomic configuration of interfaces in metallic materials." *Materials Science Forum* 294-296. (1999). 17-26. https://doi.org/10.4028/www.scientific.net/MSF.294-296.17

[77]   Needels, M., A.M. Rappe, P.D. Bristowe, and J.D. Joannopoulos. "Ab initio study of a grain boundary in gold." *Physical Review B* 46. no. 15 (1992). 9768-9771. https://doi.org/10.1103/PhysRevB.46.9768

[78]   Arias, T.A., and J.D. Joannopoulos. "Electron trapping and impurity segregation without defects: Ab initio study of perfectly rebonded grain boundaries." *Physical Review B* 49. no. 7 (1994). 4525-4531. https://doi.org/10.1103/PhysRevB.49.4525

[79]   Starostenkov, M.D., N.B. Kholodova, G.M. Poletaev, G.V. Popova, N.F. Denisova and I.A. Demina. "Computer modelling of structural-energy transformations in nanocrystals and low-dimensional systems." *Polzunovsky Almanac*. no. 4 (2003). 115-117, [in Russian].

[80]   Daw, M.S., and M.I. Baskes. "Embedded-atom method: Derivation and application to impurities and other defects in metals." *Physical Review B* 29. no. 12 (1984). 6443-6453. https://doi.org/10.1103/PhysRevB.29.6443

[81]   Foiles, S.M., M.I. Baskes, and M.S. Daw. "Embedded-atom-method functions for the fcc metals Cu, Ag, Au, Ni, Pd, Pt, and their alloys." *Physical Review B* 33. no. 12 (1986). 7983-7991. https://doi.org/10.1103/PhysRevB.33.7983

[82]   Johnson, R.A. "Analytic nearest-neighbor model for fcc metals." *Physical Review B* 37. no. 8 (1988). 3924-3931. https://doi.org/10.1103/PhysRevB.37.3924

[83]   Hijazi, I.A., and Y.H. Park. "Consistent analytic embedded atom potential for face-centered cubic metals and alloys." *Journal of Materials Science & Technology* 25. no. 6 (2009). 835-846.

[84] Johnson, R.A. "Alloy models with the embedded-atom method." *Physical Review B* 39. no. 17 (1989). 12554-12559. https://doi.org/10.1103/PhysRevB.39.12554

[85] Berendsen, H.J.C., J.P.M. Postma, W.F. van Gunsteren, A. DiNola, and J.R. Haak. "Molecular-dynamics with coupling to an external bath." *Journal of Chemical Physics* 81. no. 8 (1984). 3684-3690. https://doi.org/10.1063/1.448118

[86] Podryga, V.O., and S.V. Polyakov. "Molecular-dynamic modelling of the process of establishment of thermodynamic equilibrium of heated nickel." *Preprint of M.V. Keldysh Institute of Applied Mathematics*. no. 41 (2014), [in Russian].

[87] XMD – Molecular Dynamics for Metals and Ceramics. Accessed August 1, 2018. http://xmd.sourceforge.net/about.html.

[88] Markidonov, A.V., A.V. Yashin, A.A. Chaplygina, and N.V. Sinitsa. *Modeling the propagation of shock waves in nanoobjects by the molecular dynamics method (WAVE)*. Certificate of state registration of the computer program No. 2013661857 of December 17, 2013.

[89] RasMol and OpenRasMol. Molecular Graphics Visualisation Tool. Accessed August 1, 2018. http://www.rasmol.org/.

[90] Johnson, R.A. "Alloy models with the embedded-atom method." *Physical Review B* 39. no. 17 (1989). 12554-12559. https://doi.org/10.1103/PhysRevB.39.12554

[91] Grigoryev, I.S., and E.Z. Meilikhov. *Physical quantities*. Moscow: Energoatomizdat, 1991, [in Russian].

[92] Mogilevskii, M.A., V.V. Efremov, and I.O. Minkin. "Behavior of the crystal lattice under strong one-dimensional compression." *Physics of Combustion and Explosion*. no. 5 (1977). 750-754, [in Russian]. https://doi.org/10.1007/BF00742223

[93] Mogilevskii, M.A., and I.O. Minkin. "Influence of point defects on one-dimensional compression of a lattice." *Physics of Burning and Explosion*. no. 5 (1978). 159-163, [in Russian]. https://doi.org/10.1007/BF00789736

[94] Aksenov, M.S., G.M. Poletaev, R.Yu. Rakitin, and M.D. Starostenkov. "Study of self-diffusion in uniaxially deformed two-dimensional crystals." *Fundamental Problems of Modern Materials Science*. no. 2 (2005). 64-67, [in Russian].

[95] Krivtsov, A.M. "Description of plastic effects in the molecular dynamics simulation of spall fracture." *Physics of the Solid Body* 46. no. 6 (2004). 1025-1030, [in Russian]. https://doi.org/10.1134/1.1767244

[96] Indenbom, V.L. "A new hypothesis on the mechanism of radiation-stimulated processes." *Journal of Experimental and Theoretical Physics* 5. no. 6 (1979). 489-492, [in Russian].

[97] Shalaev, A.M. *Radiation-stimulated diffusion in metals.* Moscow: Atomizdat, 1972, [in Russain].

[98] Stepanov, V.A. "Radiation-stimulated diffusion in solids." *Journal of Technical Physics* 68. no. 8 (1998). 67-72, [in Russian]. https://doi.org/10.1134/1.1259104

[99] Poletaev, G.M., and M.D. Starostenkov. "Contributions of various mechanisms of self-diffusion in fcc metals under equilibrium conditions." *Physics of the Solid Body* 52. no. 6 (2010). 1075-1082, [in Russian]. https://doi.org/10.1134/S1063783410060065

[100] Orlov, A.N, and Yu.V. Trushin. *Energy of point defects in metals.* Moscow: Energoatomizdat, 1983, [in Russian].

[101] Bokshtein, B.S., S.Z. Bokshtein, and A.A. Zhukhovitsky. *Thermodynamics and kinetics of diffusion in solids.* Moscow: Metallurgy, 1974, [in Russian].

[102] Cherne, F.J., and M.I. Baskes. "Properties of liquid nickel: A critical comparison of EAM and MEAM calculations." *Physical Review B* 65. no. 2 (2001). https://doi.org/10.1103/PhysRevB.65.024209

[103] Gieb, M., J. Heieck, and W. Shüle. "Radiation-enhanced diffusion in nickel-10.6% chromium alloys." *Journal of Nuclear Materials* 225. (1995). 85-96. https://doi.org/10.1016/0022-3115(94)00689-X

[104] Orlov, V.L., A.V. Orlov, A.Kh. Al-Samavi, and V.V. Evstigneev. "Formation of a germ of a radiation pore." *Izvestiya VUZov. Physics* 47. no. 3 (2004). 25-28, [in Russian]. https://doi.org/10.1023/B:RUPJ.0000038741.89237.98

[105] Orlov, V.L., A.V. Orlov, A.Kh. Al-Samavi, and A.A. Grebenkov. "The temperature interval of radiation swelling." *Izvestiya VUZov. Physics* 47. no. 6 (2004). 27-30, [in Russian]. https://doi.org/10.1023/B:RUPJ.0000047841.60538.e1

[106] Aksenov, M.S., G.M. Poletaev, R.Yu. Rakitin, V.Yu. Krasnov, and M.D. Starostenkov. "Stability of vacancy clusters in fcc metals." *Fundamental Problems of Modern Materials Science* 2. no. 4 (2005). 24-31, [in Russian].

[107] Orlov, A.V., L.V. Lyskov, and V.L. Orlov. "Conditions of ascending diffusion in concentrated alloys." *Bulletin of Ugra State University*. no. 1 (2008). 102-105, [in Russian].

[108] Poletaev, G.M. (2008) Atomic mechanisms of structural-energy transformations in the volume of crystals and near the boundaries of tilt grains in fcc metals (Doctoral dissertation). Retrieved from http:// http://vak1.ed.gov.ru/common/img/uploaded/files/vak/announcements/fiz_mat/ 13-10-2008/PoletaevGM.pdf, [in Russian].

[109] Rakitin, R.Yu. (2006) Investigation of the diffusion mechanisms along the tilt grain boundaries in fcc metals (Doctoral dissertation). Retrieved from http:// http://www.nsmds.ru/files/aftoref/rakitin.pdf, [in Russian].

[110] Orlov, A.N., V.N. Perevezentsev, and Rybin V.V. *Boundaries of grains in metals*. Moscow: Metallurgy, 1980, [in Russian].

[111] Chen, S.P., D.J. Srolovitz, and Voter A.F. "Computer simulation on surfaces and [001] symmetric tilt grain boundaries in Ni, Al, and Ni3Al." *Journal of Materials Research* 4. no. 1 (1989). 62-77. https://doi.org/10.1557/JMR.1989.0062

[112] Psakhie, S.G., K.P. Zolnikov, K.I. Kadyrov, G.E. Rudensky, Yu.P. Sharkeev, and V.M. Kuznetsov. "On the possibility of forming soliton-shaped pulses during ion implantation." *Letters to the Journal of Technical Physics* 25. no. 6 (1999). 7-12, [in Russian]. https://doi.org/10.1134/1.1262425

[113] Ovidko, I.A., and A.B. Reizis. "Climbing of grain boundary dislocations and diffusion in nanocrystalline materials." *Physics of the Solid Body* 43. no. 1 (2001). 35-38, [in Russian]. https://doi.org/10.1134/1.1340182

[114] Bobylev, S.V. "Theoretical models of the emission of dislocations by grain boundaries in deformable nanocrystalline materials." *Materials Physics and Mechanics* 12. no. 2 (2011). 126-160, [in Russian].

[115] Howe, L.M. "Electron displacement damage in cobalt in a high voltage electron microscope." *Philosophical Magazine* 22. no. 179 (1970). 965-981.

[116] Novikov, I.I. *Defects of the crystal structure of metals*. Moscow: Metallurgy, 1983, [in Rusiian]. https://doi.org/10.1080/14786437008221067

[117] Kalongji, G., and J.W. Cahn. "The stacking-fault tetrahedron."
Philosophical Magazine A 53. no. 54 (1986). 521-529.
https://doi.org/10.1080/01418618608242850

[118] Khonikomb, R. *Plastic deformation of metals.* Moscow: Mir, 1972, [in
Russian].

[119] Kalin, B.A. *Physical Materials Science.* Moscow: Moscow Engineering
Physics Institute, 2007, [in Russian].

[120] Begrambekov, L.B. "Erosion and transformation of a surface under ion
bombardment." *The Achievements of Science and Technology* 7. (1993). 4-57,
[in Russian].

[121] Devyatko, Yu.N., M.Yu. Kagan, and O.V. Khomyakov. "New mechanism of
vacancy pore formation." *Low-Temperature Physics* 36. no. 4 (2010). 398-402,
[in Russian]. https://doi.org/10.1063/1.3388823

[122] Rogozhina, T.S. "Contact energy in the metal adhesion zone." *Investigated
in Russia.* (2007) Accessed August 1, 2018. http://
zhurnal.ape.relarn.ru/articles/2007/174.pdf, [in Russian].

[123] Personick, M.L., M.R. Langille, J. Zhang, and C.A. Mirkin. "Shape control
of gold nanoparticles by silver underpotential deposition." *Nano Letters* 11. no.
8 (2011). 3394-3398. https://doi.org/10.1021/nl201796s

[124] Gladkikh, N.T., S.V. Dukarov, A.P. Kryshtal, V.I. Larin, V.N. Sukhov, and
S.I. Bogatyrenko. *Surface phenomena and phase transformations in condensed
films.* Kharkiv: N.V. Karamzin Kharkiv State University, 2004, [in Russian].

[125] Magomedov, M.N. "On the dependence of surface energy on the size and
shape of a nanocrystal." *Physics of the Solid Body* 46. no. 5 (2004). 924-937, [in
Russian]. https://doi.org/10.1134/1.1744976

[126] Bekman, I.N. *The nuclear industry.* Moscow: Moscow State University
Publishing House, 2005, [in Russian].

[127] Geguzin, Ya.E. *Physics of sintering.* Moscow: Nauka, 1984, [in Russian].

[128] Bataykina, I.A., and N.P. Tikhonova. "Computer simulation of relaxed grain
boundaries and the calculation of the energy of boundaries." Proceedings of the
IV International Scientific and Practical Conference "Modern Information
Technologies and IT Education", Moscow, December 2009. Accessed August
1, 2018. http://2009.it-edu.ru/docs/Sekziya_6
/22_Butaykina_Tihonova_1256805859722725.doc, [in Russian].

[129] Sinyayev, D.V. (2008) Investigation of the mechanisms of structural-energy transformations near the tilt grain boundaries in Ni3Al intermetallic compound (Doctoral dissertation). Retrieved from http:// http://www.nsmds.ru/files/aftoref/sinyaev.pdf, [in Russian].

[130] Komarov, F.F. "Defect and track formation in solids under irradiation by ions of ultra-high energies." *Achievements of the Physical Sciences* 173. no. 12 (2003). 1287-1318, [in Russian]. https://doi.org/10.1070/PU2003v046n12ABEH001286

[131] Baranov, Y.A., Yu.V. Martynenko, S.O. Tsepelevich, and Yu.N. Yavlinsky. "Inelastic sputtering of solids by ions." *Achievements of the Physical Sciences* 156. no. 3 (1988). 477-511, [in Russian].

[132] Price, P.B., and R.M. Walker. "Electron microscope observation of etched tracks from spallation recoils in mica." *Physical Review Letters* 8. no. 5 (1962). 217-219. https://doi.org/10.1103/PhysRevLett.8.217

[133] Tretyakova, S.P. "Dielectric detectors and their use in experimental nuclear physics." *Physics of Elementary Particles and the Atomic Nucleus* 23. no. 2 (1992). 364-429, [in Russian].

[134] Houpert, Ch., F. Studer, and D. Groult. "Transition from localized defects to continuous latent tracks in magnetic insulators irradiated by high energy heavy ions: A HREM investigation." *Nuclear Instruments and Methods in Physics. Research Section B: Beam Interactions with Materials and Atoms* 39. no. 1-4 (1989). 720-723. https://doi.org/10.1016/0168-583X(89)90882-3

[135] Prokofiev, M.A., D.G. Berdonosova, I.V. Melikhov, and S.S. Berdonosov. "On the possibility of obtaining crystalline materials containing extended cylindrical pores." *Bulletin of Moscow University. Series: "Chemistry"* 51. no. 4 (2010). 325-330, [in Russian]. https://doi.org/10.3103/S0027131410040115

[136] Shiller, Z., U. Gaizig, and Z. Panzer. *Electron-beam technology*. Moscow: Energia, 1980, [in Russian].

[137] Tsvetaeva, A.A. *Defects in hardened metals*. Moscow: Atomizdat, 1969, [in Russian]

[138] Markidonov, A.V. "Influence of post-cascade shock waves on small vacancy clusters." Proceedings of IX Russian Annual Conference of Young Researchers and Graduate Students "Physical Chemistry and Technology of Inorganic Materials", Moscow, October 2012, [in Russian].

[139] Markidonov, A.V. "Influence of shock waves on vacancy pores." Proceedings of the All-Russian Youth Scientific School "Chemistry and Technology of Polymer and Composite Materials", Moscow, November 2012, [in Russian].

[140] Markidonov, A.V., M.D. Starostenkov, and E.P. Pavlovskaya. "Influence of post-cascade shock waves on processes of enlargement of vacancy pores." *Fundamental Problems of Modern Materials Science* 9. no. 4/2 (2012). 694-702, [in Russian].

[141] Markidonov, A.V., and M.D. Starostenkov. "Influence of post-cascade shock waves on small vacancy clusters." Proceedings of Scientific Readings in Memory of Corresponded Member of the Russian Academy of Sciences I.A. Oding: "Mechanical Properties of Modern Structural Materials." Moscow, September 2012, [in Russian].

[142] Markidonov, A.V., M.D. Starostenkov, and A.V. Yashin. "Structural transformations of the vacancy pore under irradiation of the material." *Fundamental Problems of Modern Materials Science* 10. (2013), 12-21, [in Russian].

[143] Markidonov, A.V., M.D. Starostenkov, E.P. Pavlovskaya, A.V. Yashin, G.M. Poletaev. "Low-temperature dissolution of pores near the surface of a crystal under the impact of shock waves." *Fundamental Problems of Modern Materials Science* 10. no. 2 (2013). 254-261, [in Russian].

[144] Markidonov, A.V., and E.P. Pavlovskaya. "Properties of structural transformations of vacancy pores under the impact of shock waves in the presence of a free crystal surface." Proceedings of the IV All-Russian Scientific and Practical Conference of Scientists, Postgraduates, Specialists and Students, with International Participation: "Modern Problems of Methodology and Innovation Activity". Novokuznetsk, May 2013, [in Russian].

[145] Starostenkov, M.D., A.V. Markidonov, and E.P. Pavlovskaya. "Structural transformations of vacancy pores under the influence of shock waves." *Bulletin of Tambov University. Series: Natural and Technical Science* 18. no. 4 (2013). 1741-1743, [in Russian].

[146] Medvedev, N.N., M.D. Starostenkov, P.V. Zakharov, and A.V. Markidonov. "Migration of aggregates of point defects in model crystals". *Bulletin of Tambov University. Series: Natural and Technical Science* 18. no. 4 (2013). 1850-1852, [in Russian].

[147] Markidonov, A.V., and E.P. Pavlovskaya. "Diffusion evolution of vacancy pores in gold caused by shock waves." Proceedings of the II International Scientific and Practical Conference "Modern Trends and Innovations in Science and Production", Mezhdurechensk, April 2013, [in Russian].

[148] Markidonov, A.V. "Enlargement of vacancy pores under the influence of shock waves." Proceedings of the XIX International Scientific and Practical Conference "Modern Technologies and Technologies". Tomsk, April 2013, [in Russian].

[149] Markidonov, A.V. "Dissolution of vacancy pores near the surface under the influence of shock waves." Proceedings of the XIX International Scientific and Practical Conference of Students, Post-graduate Students and Young Scientists "Modern Techniques and Technologies". Tomsk, April 2013, [in Russian].

[150] Markidonov, A.V., M.D. Starostenkov, E.P. Pavlovskaya, A.V. Yashin, N.N. Medvedev, P.V. Zakharov, and A.A. Sitnikov. "Splitting of the vacancy pore in the grain boundary region by a shock post-cascade wave." *Fundamental Problems of Modern Materials Science* 10. no. 3 (2013). 443-453, [in Russian].

[151] Markidonov, A.V., and E.P. Pavlovskaya. "On the possibility of dissolution of the vacancy pore under the influence of shock waves near the surface of the crystal." Proceedings of the II All-Russian Internet Conference "The Boundaries of Science 2013". Kazan, May 2013, [in Russian].

[152] Markidonov, A.V. "Computer simulation of the effect of elastic waves on a pore in a crystal." Proceedings of the All-Russian Youth Scientific Conference with International Participation "Innovations in Materials Science". Moscow, June 2013, [in Russian].

[153] Markidonov, A.V., M.D. Starostenkov, and E.P. Pavlovskaya. "Influence of post-cascade shock waves on structural transformations of the vacancy pores." *Chemical Physics and Mesoscopy* 15. no. 3 (2013). 370-377, [in Russian].

[154] Markidonov, A.V., and E.P. Pavlovskaya. "Influence of grain boundary dislocations on the dynamics of vacancy pores in the grain boundary region." Proceedings of the V International Scientific and Innovative Youth Conference "Modern Solid-State Technologies: Theory, Practice and Innovative Management". Tambov, October 2013, [in Russian].

[155] Markidonov, A.V., M.D. Starostenkov, E.P. Pavlovskaya, A.V. Yashin, N.N. Medvedev, and P.V. Zakharov. "Structural transformation of vacancy pores in a deformed crystal under the influence of shock waves." *Fundamental Problems of Modern Materials Science* 10. no. 4 (2013). 563-572, [in Russian].

[156] Markidonov, A.V., and E.P. Pavlovskaya. "On the possibility of moving the pore through the grain boundary region." Proceedings of the XIV International Scientific and Technical Ural School-Seminar of Young Metal Scientists. Ekaterinburg, November 2013, [in Russian].

[157] Markidonov, A.V., and E.P. Pavlovskaya. "On the possibility of a homogeneous growth of the vacancy pore under the influence of shock waves." Proceedings of the VIII All-Russian Conference of Young Scientists "Nanoelectronics, Nanophotonics and Nonlinear Physics". Saratov, September 2013, [in Russian].

[158] Markidonov, A.V., and E.P. Pavlovskaya. "Impact of shock waves on the vacancy pores in a deformed crystal." Proceedings of the International Scientific and Practical Conference "Creation of Highly Effective Production Facilities at Enterprises of the Mining and Metallurgical Industry". Ekaterinburg, September 2013, [in Russian].

[159] Markidonov, A.V., M.D. Starostenkov, and E.P. Pavlovskaya. "Properties of the influence of shock waves on vacancy pores in a deformed crystal." Proceedings of the V International Conference "Deformation and Destruction of Materials and Nanomaterials". Moscow, November 2013, [in Russian].

[160] Markidonov, A.V. "Dynamics mechanism of the dissolution of a vacancy pore." Proceedings of the X Russian Annual Conference of Young Researchers and Graduate Students "Physical Chemistry and Technology of Inorganic Materials." Moscow, October 2013, [in Russian].

[161] Markidonov, A.V., M.D. Starostenkov, and P.Y. Tabakov. "Splitting vacancy voids in the grain boundary region by a post-cascade shock wave." *Materials Physics and Mechanics* 18. no. 2 (2013). 148-155.

[162] Markidonov, A.V. "On the possibility of creating capillary structures in a crystal by dividing the latent tracks by shock waves (computer simulation)." *Bulletin of Kuzbass State Technical University*. no. 1 (2014). 99-103, [in Russian].

[163] Markidonov, A.V., M.D. Starostenkov, A.V. Yashin, and P.V. Zakharov. "Study of structural transformations of the pores of a cylindrical shape by a molecular dynamics method." *Fundamental Problems of Modern Materials Science* 11. no. 2 (2014). 163-173, [in Russian].

[164] Starostenkov, M.D., A.V. Markidonov, E.P. Pavlovskaya, A.V. Yashin, N.N. Medvedev, and P.V. Zakharov. "Properties of structural transformations of vacancy pores in a deformed crystal under the influence of shock waves."

Proceedings of the 1st Russian-Kazakhstan Youth Scientific and Technical Conference. Barnaul, November 2013, [in Russian].

[165] Markidonov, A.V., and M.D. Starostenkov. "Modification of porous structures by post-cascade shock waves." Abstract. The XLIV International Tulin Conference on the Physics of Interaction of Charged Particles with Crystals. Moscow, May 2014, [in Russian].

[166] Markidonov, A.V. "On structural transformations of the pores during radiation treatment of the material." Proceedings of the V International Scientific and Practical Conference "Innovative Technologies and Economics in Mechanical Engineering". Yurga, May 2014, [in Russian].
https://doi.org/10.4028/www.scientific.net/AMM.682.25

[167] Markidonov, A.V., and M.D. Starostenkov. *Radiation-dynamic processes in fcc crystals, accompanied by high-speed mass transfer.* Kemerovo: Kuzbassvuzizdat, 2014, [in Russian].

[168] Markidonov, A.V. "The coefficient of dynamic self-diffusion and its calculation using the example of nickel crystallite." Proceedings of the XI Russian Annual Conference of Young Researchers and Graduate Students "Physical Chemistry and Technology of Inorganic Materials." Moscow, September 2014, [in Russian].

[169] Markidonov, A.V. "The mechanism of self-diffusion in an fcc crystal, realized under the influence of shock waves." Proceedings of Scientific Readings in Memory of Corresponded Member of Russian Academy of Sciences I.A. Oding: "Mechanical Properties of Modern Structural Materials." Moscow, September 2014, [in Russian].

[170] Markidonov, A.V. "Calculation of the coefficient of dynamic self-diffusion by a molecular dynamics method." Proceedings of the 3rd All-Russian Internet Conference "The Boundaries of Science 2014". Kazan, May 2014, [in Russian].

[171] Markidonov, A.V., M.D. Starostenkov, P.V. Zakharov, and Z. Zhilian. "Role of post-cascade shock waves in the low-temperature activation of self-diffusion." *Fundamental Problems of Modern Materials Science* 11. no. 3 (2014). 346-353, [in Russian].

[172] Markidonov, A.V. "About the structural transformation of void during radiation treatment of the material." *Applied Mechanics and Materials* 682. (2014). 25-31. https://doi.org/10.4028/www.scientific.net/AMM.682.25

[173] Markidonov, A.V., and M.D. Starostenkov. "The process of self-diffusion in an fcc crystal caused by the passing of a shock wave." Abstract. The Open

School-Conference of the CIS Countries "Ultrafine-Grained and Nanostructured Materials". Ufa, October 2014, [in Russian].

[174] Markidonov, A.V., and M.D. Starostenkov. "Computer simulation of the fission of latent tracks in an fcc crystal by shock waves with the subsequent formation of a capillary structure." Proceedings of the VIII International Conference "Phase Transformations and Strength of Crystals". Chernogolovka, October 2014, [in Russian].

[175] Markidonov, A.V. "Activation of self-diffusion by shock waves in crystalline nickel." Proceedings of the IX All-Russian Conference of Young Scientists "Nanoelectronics, Nanophotonics and Nonlinear Physics". Saratov, September 2014, [in Russian].

[176] Markidonov, A.V., and M.V. Smirnova. "Calculation of the coefficient of dynamic self-diffusion by a computer simulation method after radiation exposure to a solid body." Proceedings of the XII All-Russian and International School-Seminar on Structural Macrokinetics for Young Scientists. Chernogolovka, November 2015, [in Russian].

[177] Markidonov, A.V., M.D. Starostenkov, A.A. Soskov, and G.M. Poletayev. "The study of structural transformations of nanopores of cylindrical shape in gold by a molecular dynamics method under the conditions of thermal activation and the action of sound and shock waves." *Physics of the Solid State* 57. no. 8 (2015). 1521-1524, [in Russian]. https://doi.org/10.1134/S106378341508017X

[178] Markidonov, A.V., M.D. Starostenkov, P.V. Zakharov, and O.V. Obidina. "Pore formation in an fcc crystal under the influence of shock post-cascade waves." *Fundamental problems of modern materials science* 12. no. 2 (2015). 231-240, [in Russian].

[179] Markidonov, A.V., M.D. Starostenkov, and G.M. Poletayev. "Transformation of nanopores in gold under conditions of thermal activation and impact of sound and shock waves." *Proceedings of the Russian Academy of Sciences. Series: Physics* 79. no. 9 (2015). 1233-1237, [in Russian]. https://doi.org/10.3103/S1062873815090130

[180] Markidonov, A.V., M.D. Starostenkov, and M.V. Smirnova. "The process of self-diffusion in an fcc crystal, caused by the passing of a shock wave." *Izvestiya VUZov. Physics* 58. no. 6 (2015). 80-84, [in Russian]. https://doi.org/10.1007/s11182-015-0576-8

[181] Markidonov, A.V., M.D. Starostenkov, A.A. Soskov, and G.M. Poletaev. "Molecular dynamics study of structural transformations of cylindrical nanopores in gold under thermal activation conditions and under the action of acoustic and shock waves." *Physics of the Solid State* 57. no. 8 (2015).1551-1554. https://doi.org/10.1134/S106378341508017X

[182] Markidonov, A.V., M.D. Starostenkov, and G.M. Poletaev. "Transformation of nanopores in gold under conditions of thermoactivation and the effects of acoustic and shock waves." *Bulletin of the Russian Academy of Sciences. Physics* 79. no. 9 (2015). 1089-1092. https://doi.org/10.3103/S1062873815090130

[183] Markidonov, A.V. "Homogeneous nucleation of pores in an fcc crystal under the influence of post-cascade shock waves." Proceedings of the XII Russian Annual Conference of Young Researchers and Graduate Students "Physical Chemistry and Technology of Inorganic Materials" (with international participation). Moscow, October 2015, [in Russian].

[184] Markidonov, A.V. "Investigation of the process of pore formation at the grain boundary in an fcc crystal under irradiation." Proceedings of the 4th All-Russian Internet Conference "The Boundaries of Science 2015". Kazan, June 2015, [in Russian].

[185] Markidonov, A.V., M.D. Starostenkov, and M.V. Smirnova. "Self-diffusion process in an fcc crystal caused by the passage of a shock wave." *Russian Physics Journal* 58. no. 6 (2015). 828-832. https://doi.org/10.1007/s11182-015-0576-8

[186] Markidonov, A.V. "Pore formation at grain boundaries under the influence of shock waves arising from radiation exposure." Proceedings of the XIII All-Russian (with the international participation) School-Seminar on Structural Macrokinetics for Young Scientists Organized in the Memory of Academician A.G. Merzhanov. Chernogolovka, November 2015, [in Russian].

[187] Markidonov, A.V., and M.D. Starostenkov. "Pore formation in the grain-boundary region of fcc crystals under the influence of post-cascade shock waves." Proceedings of the VI International Conference "Deformation and Destruction of Materials and Nanomaterials". Moscow. May 2015, [in Russian].

[188] Markidonov, A.V., and M.D. Starostenkov. "Coalescence of vacancy nanopores in a crystal of an fcc lattice under the influence of post-cascade shock waves." *Bulletin of Voronezh State University. Series: "Physics. Mathematics".* no. 1 (2016). 14-23, [in Russian].

[189] Markidonov, A.V., and M.D. Starostenkov. "On the possibility of homogeneous nucleation of a pore in a grain boundary region under the influence of post-cascade shock waves." *Problems of Atomic Science and Technology. Series: "Mathematical Modelling of Physical Processes".* no. 3 (2016). 37-46, [in Russian].

[190] Bykov, A.A. *Development of manufacturing of bimetals.* Metallurg: Scientific, Technical and Production Magazine. 2009, [in Russian].

[191] Korol, V.K., and M.S. Gildengorn. *Foundations of technology for production of multilayer metals.* Moscow: Metallurgy. 1970, [in Russian].

[192] Pavlov, I.N., V.N. Lebedev, A.G. Kobelev et al. *Layered compositions.* Moscow: Metallurgy. 1986, [in Russian].

[193] Bondarchuk, I.S., and F.J.C.S. Aires. "Bimetallic supported Pd-Ag catalysts." Proceedings of the XI International Conference of Students and Young Scientists "Prospects for the Development of Basic Science". Tomsk. April 2014, [in Russian].

[194] Sanyal, U., D.T. Davis, and B.R. Jagirdar. "Bimetallic core-shell nanocomposites using weak reducing agent and their transformation to alloy nanostructures." *Dalton Trans* 42. no. 19 (2013). 7147-7157. https://doi.org/10.1039/c3dt33086a

[195] Eliana Misi, S.E., A. Ramli, and F.H. Rahman. "Characterization of the structure feature of bimetallic Fe-Ni catalysts." *Journal of Applied Science* 11. (2011). 1297-1302. https://doi.org/10.3923/jas.2011.1297.1302

[196] Asadullah, M., T. Miyazawa, S. Ito, K. Kunimori, and K. Tomishige. "Demonstration of real biomass gasification drastically promoted by effective catalyst." *Applied Catalysis A: Gen* 246. (2013). 103-116. https://doi.org/10.1016/S0926-860X(03)00047-4

[197] Tian, D., Z. Liu, D. Li, H. Shi, W. Pan, and Y. Cheng. "Bimetallic Ni-Fe total-methanation catalyst for the production of substitute natural gas under high pressure." *Fuel* 104. (2013). 224-229. https://doi.org/10.1016/j.fuel.2012.08.033

[198] Shen, Y., W. Gong, B. Zheng, and L. Gao. "Ni-Al bimetallic catalysts for the preparation of multiwalled carbon nanotubes from polypropylene: Influence of the Ni-Al ratio." *Applied Catalysis B: Environmental* 181. (2016). 769-778. https://doi.org/10.1016/j.apcatb.2015.08.051

[199] Massicot, F., R. Schneider, Y. Fort, S. Illy-Cherrey, and O. Tillement. "Synergistic effect in bimetallic Ni-Al clusters. Application to efficient catalytic reductive dehalogenation of polychlorinated arenas." *Tetrahedron* 6. no. 27 (2000). 4765-4768. https://doi.org/10.1016/S0040-4020(00)00383-5

[200] Poletaev, G.M. "Modelling by a molecular dynamics method of structural-energy transformations in two-dimensional metals and alloys (MD2)." ROSPATENT Certificate No. 2008610486, January 25, 2008, [in Russian].

[201] Tsaregorodtsev, A.I., N.V. Gorlov, B.F. Demyanov, and M.D. Starostenkov. "The atomic structure of the antiphase boundaries and its effect on the state of the lattice near the dislocation in ordered alloys with the $L1_2$ superstructure". *Physics of Metals and Metallurgy* 58. no. 2 (1984). 336-343, [in Russian].

[202] Poletaev, G.M. (2002) Investigation of mutual diffusion processes in a two-dimensional system of Ni-Al (Doctoral dissertation). Retrieved from: https://elibrary.ru/item.asp?id=25238860, [in Russian].

[203] Starostenkov, M.D., P.V. Zakharov, N.N. Medvedev, A.V. Markidonov, A.M. Eremin, A.A. Soskov, and V.R. Mikryukov. "Features of the process of mass transfer in various bimetals in the presence of vacancy complexes in the dislocation field of discrepancies." *Fundamental Problems of Modern Materials Science* 10 no. 2 (2013). 245-250, [in Russian].

[204] Zakharov, P.V., M.D. Starostenkov, N.N. Medvedev, A.V. Markidonov, and A.M. Eremin. "The role of vacancy complexes in the dislocation field of discrepancies during mass transfer at the interphase boundary of bimetals." Collection of scientific articles of the international youth school-seminar "Lomonosov Readings in the Altai", Barnaul, November 2013, [in Russian].

[205] Zakharov, P.V., M.D. Starostenkov, and A.V. Markidonov. "Modelling of cooperative processes at the boundary of Pt-Al bimetal in the presence of defects." Proceedings of the XIV International Scientific and Technical Ural School-Seminar of Young Metal Scientists. Ekaterinburg, November 2013, [in Russian].

[206] Zakharov, P.V., M.D. Starostenkov, N.N. Medvedev, A.V. Markidonov, and O.V. Obidina. "Cooperative behaviour of interstitial atoms in the field of misfit dislocations at the Ni-Al bimetal boundary." *Fundamental Problems of Modern Materials Science* 9. no. 4 (2012). 431-435, [in Russian].

[207] Zakharov, P.V., N.N. Medvedev, and M.D. Starostenkov. "Effects of self-organization of matter at the atomic level during passage of a solitary transverse

wave across the Ni-Al bimetal boundary." *Fundamental Problems of Modern Materials Science* 9. no. 1 (2012). 46-49, [in Russian].

[208] Starostenkov, M.D., P.V. Zakharov, and N.N. Medvedev. "Interaction of crowdion with the Ni-Al bimetal boundary in a 2D model." *Letters on Materials* 1. no. 4 (2011). 238-240, [in Russian]. https://doi.org/10.22226/2410-3535-2011-4-238-240

[209] Koshkin, N.I., and M.G. Shirkevich. *Handbook of elementary physics*. Moscow: Science. 1972, [in Russian].

Keyword index

About the authors

Mikhail D. Starostenkov, Dr. Sci. (Phys.-math), Full Professor, Head of the Department of Physics, Federal State Budget Educational Institution of Higher Education "Polzunov Altai State Technical University", Barnaul, Russia. He is the author of more than 1,500 articles in the field of materials science. He is the founder of the scientific school "Theory and Computer Modeling in Condensed Matter Physics". The scientific interests of Mikhail D. Starostenkov are fundamental questions of materials science and condensed matter physics. The main direction is the study of the evolution of defective structures in metals and alloys by computer simulation.

Artem V. Markidonov, Dr. Sci. (Phys.-math), Assoc. Professor, Head of the Department of Informatics and Computer Engineering named after V.K. Butorin, Novokuznetsk Institute (branch) Kemerovo State University, Russia. He is the author of more than 100 articles in the field of computer simulation of materials. His research interests are modeling structural changes and defect formation that occur in crystals under external high-energy influences.

Pavel V. Zakharov, Dr. Sci. (Phys.-math), Assoc. Professor, Professor of the Department of Mathematics, Physics, Computer Science, Federal State Budget Educational Institution of Higher Education "Shukshin Altai State University for Humanities and Pedagogy", Biysk, Russia. He is the author of more than 100 articles in the field of computer simulation of materials. His research interests include modeling structural changes in crystals under the influence of high-energy influences. As well as the study of the localization of energy and soliton-like waves.

Pavel Y. Tabakov graduated from the Kiev Institute of Civil Engineering, Ukraine, in 1986. He obtained his Ph.D. from the Department of Mechanical Engineering at the University of Natal, South Africa. He is currently a Full Professor in the Department of Mechanical Engineering at Durban University of Technology, South Africa. His areas of research include structural analysis, anisotropic materials and nanocomposites, design optimization analysis of engineering structures using artificial intelligence and evolutionary algorithms. In addition, he is interested in classifying and clustering data into large, complex multidimensional databases. He is author and co-author of more than eighty articles in international journals and conferences.

www.ingramcontent.com/pod-product-compliance
Lightning Source LLC
Chambersburg PA
CBHW071708210326
41597CB00017B/2383